# FROM
# ALGEBRA
# TO **COMPUTATIONAL**
# **ALGORITHMS:**
## Kolmogorov and Hilbert's Problem 13

## David A. Sprecher

Docent
Press

DOCENT PRESS
Boston, Massachusetts, USA
www.docentpress.com

Docent Press publishes books in the history of mathematics and computing about interesting people and intriguing ideas. The histories are told at many levels of detail and depth that can be explored at leisure by the general reader.

Cover design by Brenda Riddell, Graphic Details.

Produced with TeX.

# Contents

# List of Figures

*Dedicated to my granddaughter Lila Augusta Fox*
*Whose soaring imagination and abounding creativity*
*Were a source of inspiration throughout this project*

"It is strange... that Hilbert formulated this problem of algebraic geometry in terms of functions of real variables – but he did."

V. I. Arnold
*Hilbert's superposition to Dynamical Systems*

"... Kolmogorov's theorem was both powerful and shocking"

Robert Hecht-Nielsen
*Kolmogorov's Mapping Neural Network Existence Theorem*

"Applications of Kolmogorov's superposition theorem to nonlinear circuit and system theory, statistical pattern recognition, and large and multidimensional signal processing are presented and discussed"

Rui J. P. De Figueiredo
*Implications and Applications of Kolmogorov's Superposition Theorem*

# Acknowledgments

Forty years ago I was asked by Richard Bellman of the University of Southern California, and independently by Gian-Carlo Rota of MIT, to expand into a book a survey article *A Survey of Solved and Unsolved Problems in Superpositions of Functions* (Sprecher [165]), but extraneous duties got in the way. The subject of Kolmogorov's function representations has expanded since because of the development of computational algorithms and its application to computing following Robert Hecht-Nielsen's interpretation of Kolmogorov's formula as a feedforward neural network. When Scott Guthery proposed this project I was not sure if I was up to reviewing well over two hundred references. He was both persuasive and persistent, and this monograph was completed and improved thanks to his mathematical insights and efforts that included cogent editorial advice and guidance.

Thanks are also due to Professor Bernard Fusaro of the University of Southern Florida for numerous discussions during various stages of writing, for insightful reviews of early drafts and editorial suggestions.

# Preface

The cynosure of this monograph is a surprising assertion about real-valued continuous functions of two or more variables.

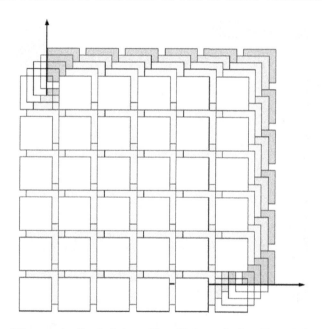

Figure 1: Satisfying Condition **A** for Fixed $k$

Consult Figure 1 and imagine that you are given...

**Condition A** An integer $n \geq 2$ and $2n + 1$ families of mutually disjoint $n$-dimensional closed cubes such that:

$$S_{q,k} \cap S_{q,k'} = \emptyset, \tag{1}$$

for $k \neq k', k = 1, 2, 3, \ldots$ and $q = 1, 2, \ldots, 2n + 1$.

The diameters of the cubes diminish to zero as $k$ tends to infinity, and such that every point of the unit cube $E^n$ belongs to at least $n + 1$ of them for each $k$.

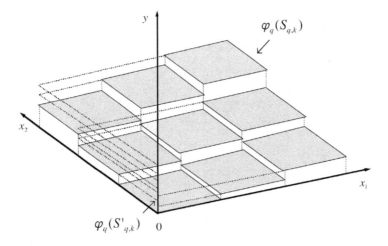

Figure 2: Satisfying Condition **B** for Fixed $k$

Now, consult Figure 2 and imagine further that you have...

**Condition B** $2n + 1$ continuous functions of $n$ variables defined on $E^n$:

$$y_q = \varphi_q(x_1, \ldots, x_n), \quad q = 1, 2, \ldots, 2n + 1,$$

with the property:

$$\varphi_q\left(S_{q,k}\right) \cap \varphi_q\left(S_{q,k'}\right) = \emptyset \tag{2}$$

whenever equation (1) is true.

Then you can represent every real-valued continuous function defined on $E^n$ in the fixed format

$$f(x_1, ..., x_n) = \sum_{q=1}^{2n+1} g_q[\varphi_q(x_1, ..., x_n)] \tag{3}$$

with continuous functions $g_q$.

To aid in absorbing this formula, think about functions of two variables defined on the unit square. For example:

$$2x_1^3 + 4x_1 x_2^2 - 5x_2 = \sum_{q=1}^{5} g_q[\varphi_q(x_1, x_2)]$$

$$x_1 \cos a x_2 = \sum_{q=1}^{5} h_q[\varphi_q(x_1, x_2)]$$

$$e^{\sqrt{x_1^2 + x_2^2}} = \sum_{q=1}^{5} l_q[\varphi_q(x_1, x_2)]$$

One format fits all!

This might not quite make sense. The five inner functions $\varphi_q(x_1, x_2)$ remain the same in each of these representations. Only the outer functions change with different target functions, and all of them are continuous.

The families of squares depicted in Figure 1 satisfy condition **A** for fixed $k$. Examine the upper left and lower right parts of the figure; you may be able to convince yourself that every point $(x_1, x_2)$ is contained in at least one of the three families corresponding to $q = 1, 2, 3$, and in at least three of the total of five families.

The next figure illustrates Property **B**, also in the case $n = 2$; each function maps disjoint cubes $S_{q,k}$ onto disjoint intervals for fixed $q$.

That each function $g_q$ in formula (3) is a function of a single variable is highlighted in the equivalent representation:

$$\begin{cases} f(x_1, ..., x_n) = \sum_{q=1}^{2n+1} g_q(y_q) \\ \\ y_q = \varphi_q(x_1, ..., x_n) \end{cases} \tag{4}$$

As astonishing as this may appear, this is only the beginning; the lesser part of a more sweeping discovery of Andrey Nikolaevich Kolmogorov. He went a significant step beyond this formula and proved remarkably that also the functions $\varphi_q$ themselves can be written as a sum of continuous functions of one variable:

$$\varphi_q(x_1, \ldots, x_n) = \psi_{1,q}(x_1) + \psi_{2,q}(x_2) + \ldots + \psi_{n,q}(x_n).$$

This allows formula (4) to be replaced by

$$\begin{cases} f(x_1, ..., x_n) = \sum_{q=1}^{2n+1} g_q(y_q) \\ \\ y_q = \psi_{1,q}(x_1) + \psi_{2,q}(x_2) + \ldots + \psi_{n,q}(x_n) \end{cases} \tag{5}$$

As far as the above examples are concerned, we can actually write:

$$2x_1^3 + 4x_1x_2^2 - 5x_2 = \sum_{q=1}^{5} g_q[\psi_{1q}(x_p) + \psi_{2q}(x_p)]$$

$$x_1 \cos ax_2 = \sum_{q=1}^{5} h_q[\psi_{1q}(x_p) + \psi_{2q}(x_p)]$$

$$e^{\sqrt{x_1^2 + x_2^2}} = \sum_{q=1}^{5} l_q[\psi_{1q}(x_p) + \psi_{2q}(x_p)].$$

If anything, this may make these representations even less believable! "*powerful and shocking*" in the words of Robert Hecht-Nielsen. This final result, formula (5), encapsulates what has become famously known as *Kolmogorov's Representation Theorem*. It is often summarized in the compact form

$$f(x_1, ..., x_n) = \sum_{q=1}^{2n+1} g_q \left[ \psi_{pq}(x_p) \right] \tag{6}$$

According to this, any continuous function of $n \geq 2$ variables defined on the $n$-dimensional Euclidean unit cube, no matter in what form it may be written or described, can be expressed in this particular format. Changing the number of variables only increases the number of summands. Nothing else changes in the format.

As daunting as formula (6) may appear, our intent is to disassemble it into its individual components in the ensuing pages, and then reassemble it step by step in computational form.

To bridge a possible gap between mathematical concepts that may be by now a fleeting memory of basic calculus, you will find from time to time thumbnail summaries of concepts pertinent to the understanding of the material. These are inserted to make the narrative as self-contained as possible.

Functions $y = f(x_1, \ldots, x_n)$ are among the most basic mathematical concepts, and continuity is one of the most intuitive concepts: a curve without breaks. We have seen how Kolmogorov's function representations relate to these heterogeneous types. They propose the idea that any continuous function of any number of variables can be composed with uniform building blocks that involve only functions of one variable and the binary operation of addition. This is a particularly impressive idea when thinking of formula (6) in the context of these examples.

Kolmogorov announced his discovery in a four-and-a-half page communication in the 1957 Number 114 issue of the Russian journal *Doklady Academii Nauk SSSR* under the title: *On the representation of continuous functions of many variables by superposition of continuous functions of one variable and addition.* It is reference [55] in the bibliography

It came on the heels of an unexpected breakthrough a year earlier, when the possibility of representing all continuous functions of several variables with continuous functions of fewer variables became a reality. It was a surprise, even to Kolmogorov. The announcement struck like lightening out of a blue sky, unanticipated and without bravado; it was a watershed discovery in the history of the concept of continuity of multidimensional real-valued functions.

Even Kolmogorov's brilliant student, Antoli Georgievich Vituškin who worked under his direction on a related problem, had no inkling two years earlier of what was coming.

Computation of a function $f(x_1, x_2)$ using Kolmogorov's formula is illustrated in Figure 3.

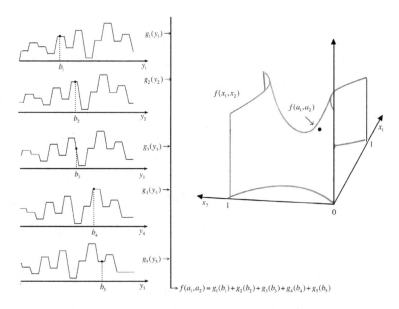

Figure 3: $f(x_1, x_2) = g_1(y_1) + g_2(y_2) + \cdots + g_5(y_5)$

Observe that a point $(x_1, x_2)$ in the plane determines five points $y_1, y_2, y_3, y_4, y_5$:

$$y_1 = \psi_{1,1}(x_1) + \psi_{2,1}(x_2)$$
$$y_2 = \psi_{1,2}(x_1) + \psi_{2,2}(x_2)$$
$$\vdots$$
$$y_5 = \psi_{1,5}(x_1) + \psi_{2,5}(x_2)$$

at which the five functions $g_1$, $g_2$, $g_3$, $g_4$, $g_5$ compute $f$. A point $(x_1, x_2)$ represents an *ordered pair* , because $x_1$ and $x_2$ represent different planar coordinates, whereas the points $y_1$, $y_2$, $y_3$, $y_4$, $y_5$ can be put in any order without changing the result. Significantly, this means that these computations can therefore be carried out in parallel, independently of each other.

The term *superpositions* in the title of Kolmogorov's announcement entered as a common proper noun the vocabulary of mathematics and computer science alike. It became synonymous with the Kolmogorov Representation Theorem that is the subject of the above-cited paper; a sobriquet was attached to it, referring to it at times as KRT; as an intimate acquaintance is referred to fondly with initials.

To clarify the terminology: a function representation in the sense of Kolmogorov is a sum of superpositions of functions of one variable; superpositions are functions of functions.

We now make a salient point with the simple product function $f(x_1, x_2) = x_1 \cdot x_2$ by writing it as a sum of two squares:

$$x_1 \cdot x_2 = \frac{1}{4}(x_1 + x_2)^2 - \frac{1}{4}(x_1 - x_2)^2.$$

Here the product of variable quantities no longer appears as such; it is replaced by the external functions

$$g_1(t) = -g_2(t) = \frac{1}{4}t^2.$$

With these, the product is expressible as the sum

$$x_1 \cdot x_2 = g_1(x_1 + x_2) + g_2(x_1 + (-x_2)).$$

Now the only arithmetic operation is addition. Risking pedantry, we point out the obvious: that the functions $g_1$ and $g_2$ are determined by the product that they actually compute.

Let's push this envelope one more notch by introducing four linear functions

$$\psi_{1,1}(x) = \psi_{2,1}(x) = \psi_{1,2}(x) = -\psi_{2,2}(x).$$

Again using only arithmetic, the product can now be represented as the sum of two superpositions of continuous functions of one variable and addition:

$$x_1 \cdot x_2 = g_1[\psi_{1,1}(x_1) + \psi_{2,1}(x_2)] + g_2[\psi_{1,2}(x_1) + \psi_{2,2}(x_2)];$$

equivalently, this can be expressed in the form

$$x_1 \cdot x_2 = \sum_{q=1}^{2} g_q[\psi_{1,q}(x_1) + \psi_{2,q}(x_2)].$$

The reader can justly complain about this complicated manipulation of mathematical symbols to express the product of two variables with superpositions of functions of one variable, but we did it for a purpose. Firstly, this is like formula (6) with two summands. Second, it is a remarkable fact that, according to Kolmogorov, there exist continuous functions $\psi_{pq}(x)$ of one variable, more complicated than the specific functions used here, such that *every* continuous function $f(x_1, x_2)$ of two variables defined on the unit square,

$$E^2 = [0, 1] \times [0, 1],$$

can be represented with five summands of this form:

$$f(x_1, x_2) = \sum_{q=1}^{5} g_q[\psi_{1,q}(x_1) + \psi_{2,q}(x_2)].$$

This is, of course, Kolmogorov's representation formula for functions of two variables.[1]

Of the several smoothness categories into which functions of any number of variables are traditionally classified in calculus and real analysis texts, continuity is the weak sister. Using the number of variables as an additional classification category is futile according to this discovery.

But Kolmogorov's discovery tells us much more: It implies that continuous functions of one variable are as convoluted as the most convoluted continuous functions of any number of variables.

Following this line of thought, think of the 'worst' continuous functions $f(x_1, ..., x_n)$ of more than one variable that you can imagine, and look at Kolmogorov's formula. Its right side says that these are no worse than the continuous functions of one variable. This runs counter to our intuition because of the expectation that an increase in the number of variables afford an increase in degrees of freedom, with an attendant increase in some measurable way of complexity in the behavior of functions.

We shall occasionally omit the qualifier *continuous* when speaking of functions, but the reader must always be careful to keep continuity in mind. We will show later that every continuous function of $n$ variables can be written as a sum on $n$ *discontinuous* functions $h_q(x)$:

$$f(x_1, ..., x_n) = \sum_{q=1}^{n} h_q(x_q).$$

To recapitulate: superpositions are functions of functions, and a function representation is a sum of superpositions. Kolmogorov's formula in the form:

$$\begin{cases} y_q = \psi_{1,q}(x_1) + \psi_{2,q}(x_2) + ... + \psi_{n,q}(x_n) \\ f(x_1, ..., x_n) = \sum_{q=1}^{2n+1} g_q(y_q) \end{cases}$$

---

[1]See Figure 3.

highlights that the computation of the functions $y_q$ is carried out before a target function $f$ is computed.

What can be simpler than these elegant function representations, composed as they are of superpositions of functions of one variable and addition? Significantly it was already indicated that the functions $\psi_{pq}(x_p)$ are completely independent of any functions $f$ being represented. They depend only on $n$ as a parameter, and once constructed can be stored and used to represent all continuous functions of $n$ variables with domain $E^n$. The single-variable functions $g_q$ that compute $f$ are continuous, so that all constructions and computations are carried out within the realm of continuous functions.

Loosely speaking, we might say that these representations imply that there are no continuous functions of more than one variable. This fundamental discovery about the concept of continuity is proved in series of inferences and logical deductions the way mathematician go about establishing propositions, but these proofs don't tell us how to compute any part of these formulas for any specific target function; they even don't tell us how to compute the functions $\psi_{pq}(x)$, but this is in the nature of mathematical existence theorems: they are of theoretical interest rather than of direct practical relevance.

The British mathematician G. H. Hardy used to boast that he never did mathematics that was useful in any way. Indeed, Kolmogorov's theorem is one of existence. Its discovery was not concerned with computation. It tells us that the functions $\psi_{pq}(x)$ and $g_q(y)$ exist according to verifiable mathematical rules.

This, of course, does not preclude the possibility that function representations *can* be computed. They have the attraction of ultimate simplicity, and the alure of a potential for the development of a computational tool—an algorithm that simplifies computation involving many variables. This is particularly attractive because the functions $\psi_{pq}(x)$ which depend only on the number of variables can be set up in advance.

Set
$$\psi_q(x) = \Sigma_{p=1}^n \psi_{pq}(x_p).$$

From the observation that the functions

$$\psi_{1,q}(x), ..., \psi_{n,q}(x)$$

determine uniquely a target function $f$, we deduce that every continuous function $f$ defined on $E^n$ is completely specified by an ordered pair $f = (g_q; \Sigma)$ in which $\Sigma$ is the rule that associates with each point $(x_1, ..., x_n)$ in $E^n$ a collection of $2n + 1$ unique sets of other points $(\psi_{1,q}(x), ..., \psi_{n,q}(x))$. Remarkably these functions $g_q$ can be permuted and the specification of $f$ is independent of the order in which they are arranged.

Yet, with all the explanations so far, formula (6) in all its manifestations is in the nature of a black box in the process of implementing function representations. This is illustrated in Figure 4 in the case of functions of two variables. The diagram shows the two stages in the computation of a function $f(x_1, x_2)$: In the first stage, the computation converting spatial variables $(x_1, x_2)$ to new variables $(y_1, y_2, y_3, y_4, y_5)$ is independent of $f$, and only in the second stage is $f$ being computed. We shall repeatedly encounter versions of this basic theme in the sequel.

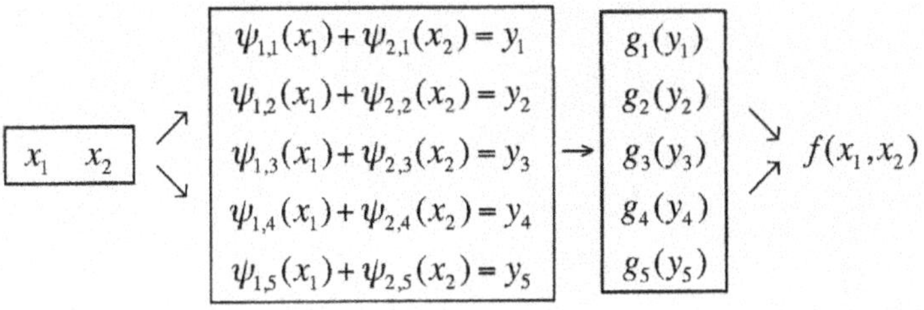

Figure 4: Black-box converting $x$ to $y$.
Schematic Representation of Kolmogorov's Formula

The hallmark of Kolmogorov's superpositions is the use of addition – the most intuitive and elementary of the four arithmetic operations. This is particularly impressive when we think of the variety of functions noted earlier. One of the goals of this monograph is to explore explanations and ramifications of having this single format represent all continuous functions of two or more variables.

We will discover subsequently that the key to this discovery was another counterintuitive concept: that of a continuous configuration that passes through every point of a square or a cube without crossing itself or having loops, called a *tree*.

For now you might think of a tree as an abstraction of the one-dimensional configuration in Figure 5 that a child might draw. Such an abstraction has infinitely many branches that fill the space. There exists a *universal tree* containing a copy of every other tree and passing through every point of a square. Think of a function whose values are arranged along the branches of the tree; this induces some order in looking up values of this function, but this gets ahead of the story

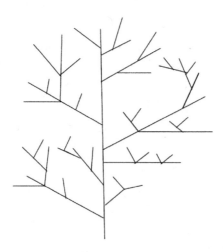

Figure 5: A Tree

The origin of Kolmogorov's discovery was an ill-fated problem, famously known as Hilbert's Problem 13. It preceded it by fifty-seven years, when David Hilbert introduced the concept of superpositions as a technique for solving a problem in algebra, making him the undisputed godfather of KRT. This will be explained more fully in the sequel.

Hilbert posed it at the International Congress of Mathematicians held in Paris in the year 1900. It was one of twenty-three problems presented by the then rising star that moved a problem about finding the roots of a polynomial equation of the seventh degree into prominence in the worldwide mathematical community. The connection between these two events: computing roots of a specific polynomial equation and the representation of arbitrary continuous functions of any number of variables constitutes part of the narrative of this monograph.

What was the problem that started it all? Recall high school algebra exercises of finding roots of quadratic equations, such as $2x^2 + 5x - 3 = 0$. These roots depend on the coefficients 2, 5, -3. Hilbert made the conjecture that the roots of the specific seventh degree polynomial equation:

$$x^7 + ax^3 + bx^2 + cx + 1 = 0, \tag{7}$$

as functions of the parameter coefficients $a$, $b$, and $c$, cannot be represented with superpositions of continuous functions of two variables (Hilbert [44]). We will have much to say about this in Chapter Five.

This explicit narrowly phrased problem in algebra does not appear at first blush to be other than routine, but then we wonder: Hilbert was one of the most illustrious mathematicians of the twentieth century. He would not single out a routine problem.

We also wonder what superpositions of continuous functions are doing here? They are not routinely involved in solving algebraic problems, such as finding roots; we learned in high school and college algebra that roots are numbers found through algebraic manipulations.

It was well known at the time that there was a range of methods for finding the roots of a specific algebraic equation. Hilbert's conjecture concerned by implications the roots of *every* polynomial equation, and not the roots of this specific equation. Readers may recall from their calculus days such a technique in which algebra and functions did meet: it was Newton's method for solving equations. This method is based on the variable $x$, and in the case at hand, it would be useful only in finding the roots $\rho(a, b, c)$ of a specific choice of coefficients $a$, $b$, and $c$.

As is generally the case with problems in mathematics, it is not the solution per se that is of interest to the mathematician; it is instead the new concepts, tools and theories that have to be developed for solving a problem. Hilbert's specific conjecture was the harbinger of an original approach that married algebra and mathematical analysis in a new way. It connected continuous functions with the concept of reduction of the number of variables in this specific situation, and this was where Kolmogorov's representations and Hilbert's problem met.

A sweeping logical extension is implied by Hilbert's problem. To see what it is, we return once more to the product function $f(x_1, x_2) = x_1 \cdot x_2$ that we represented with superpositions of functions of one variable. The connection between the number two of variables and the representation with superpositions of functions of one variable can be quantified with the expression $m_f(2) = 1$, read:

> The function $f$ of 2 variables is represented with superpositions of functions of 1 variable.

With this convention, Hilbert's conjecture for the roots $\rho(a, b, c)$ can be expressed by writing $m_\rho(3) > 2$. Namely: this function of three variables requires for its representation superpositions of functions of more than two variables. Alternatively, Hilbert's assertion implies that $m_\rho(3) = 3$; that is, the function $\rho(a, b, c)$ is a specific example of a function whose representation requires three variables.

The function

$$G(a, b, c) = g_1[h_1(a, b), h_2(a, c)] + g_2[h_3(b, c), h_4(a)]$$

is an example of a function of class $m_\rho(3) = 2$.

Using the yardstick of representations with superpositions suggests the possibility that for every integer $n \geq 3$ there is a continuous function $f(x_1, \ldots, x_n)$ for which $m_f(n) = n$. That is: For every integer $n \geq 3$ there is a continuous function $f(x_1, \ldots, x_n)$ not representable with superpositions of continuous functions of less than $n$ variables. This implies that with this yardstick, the number of variables might be a useful classification index for continuous functions.

Hilbert's conjecture was phrased for $n = 3$. For his purpose, he needed only a specific continuous function to establish the validity of the principle of using the number of variables as a classification index, and he thought that a root $\rho(a, b, c)$ of the seventh degree polynomial equation (7) provided an example of an appropriate function. We shall see later that Hilbert displayed shrewd economy by using the most elementary case to introduce a general principle tying continuity to dimension.

Hilbert's Problem 13 led to two separate lines of research with a common goal but divergent methodologies: One was aimed at validating directly the conjecture in its algebraic setting. This approach can be found in papers of Bieberbach [11-14], Otsrovski [77], Polyá and Szegö [79], Raudernbush [80], and also in a 1927 paper of Hilbert himself when he returned to the subject of his conjecture in a wider algebraic setting (Hilbert [45]).

It is important to note that in these early studies, superpositions as tools for representations of functions have not been an object of study. Instead, their role was limited to an alternative method for failed algebraic manipulations.

The other line of research was an investigation into representations of functions in general. Bari's 1930 work [8] was an early example of this, and so was Kronrod's 1980 paper [58]. These were important studies relating to what followed.

The parting of the ways eluded to above was evident in the contrast with Hilbert's conception in which Kolmogorov formulated a different general problem: Exploring the existence of dimension-reducing superpositions for representing continuous functions $f(x_1, \ldots, x_n)$ such that $m_f(n) < n$.

The ultimate coup de grâce to Hilbert's conception was Kolmogorov's famous superposition formula (6) that showed that

$$m_f(n) = 1$$

for all continuous functions $f(x_1, \ldots, x_n)$ regardless of $n$. Kolmogorov could not avail himself of the simplicity of Hilbert's approach, and he could only make his case for all continuous functions. With this approach Hilbert's conjecture was destined to become a corollary of representations with superpositions.

The importance of Kolmogorov's approach to what might be loosely termed *functional complexity* cannot be overstated: Kolmogorov set out to study function representations such that for every integer $n$ there was an integer $m \leqslant n$ for which $m_f(n) = m$. A reading of a 1956 paper of his shows that he discovered that $m_f(n) = 3$ when $n \geqslant 3$ before his final superposition theorem a year later.

Setting chronology aside, the narrative begins with the spectacular function representations of Kolmogorov. This discovery and Hilbert's problem that started it are bridged by a story that had two protagonists: Kolmogorov and his then bright nineteen-year old student Vladimir Igorovich Arnold.

We begin our story with Kolmogorov's attempt at dimension-reducing representations in the 1956 paper that we just mentioned.

We shall see that this work established the strategy for obtaining dimension-reducing superpositions, but it fell short by a hair's breadth from resolving Hilbert's conjecture due its specificity: Kolmogorov was able to show that $m_f(n) = 3$ for all continuous functions $f(x_1, ..., x_n)$ when $n \geqslant 3$, but he was not able to decide if $m_f(3) = 2$ was possible. Remember, however, that Hilbert's conjecture specifically named superpositions with two variables.

This strategy was based on the concept of *tree* that was briefly introduced earlier. More specifically, it was based on the concept of *universal tree* discovered by Menger [70] and elaborated on later in the text. It resulted in representations with a dramatic breakthrough that Kolmogorov termed the *"rather unexpected consequence"* that $m_f(n) = 3$. This meant that the number of variables $n$ was not a meaningful classification index for continuous functions of more than three variables.

Arnold refined Kolmogorov's work with shrewd insight, extending it to representations with superpositions of functions of two variables. Specifically, he proved the existence of nine real-valued continuous functions of two variables:

$$y_1 = \varphi_1(x_1, x_2)$$
$$y_2 = \varphi_2(x_1, x_2)$$
$$\vdots$$
$$y_9 = \varphi_9(x_1, x_2)$$

with which any continuous function $f(x_1, x_2, x_3)$ of three variables defined on the unit cube $E^3$ had a representation:

$$f(x_1, x_2, x_3) = g_1(y_1, x_3) + g_2(y_2, x_3) + ... + g_9(y_9, x_3)$$

where the functions $g_q$ that compute $f$ are continuous (Arnold [3]).

Keep in mind that this result was sandwiched in the short time interval between Kolmogorov's 1956 paper and his final 1957 function representation theorem. It is reasonable to assume that Kolmogorov himself might have likely taken the next step in disproving Hilbert's conjecture, but it was Arnold who did it with a short gestation period and with apparently no prior exposure to this area of mathematics. Quite a feat for a nineteen-year old student!

Here and throughout this monograph we often deviate from the common compact presentation of superposition formulas such as

$$f(x_1, x_2, x_3) = \sum_{q=1}^{9} g_q[\varphi_q(x_1, x_2), x_3]$$

We do this to highlight the inner functions that determine the characteristics of the function representation in their role of transforming the domains in which computations of a target function are performed.

We will interpret this formula later. For now we note that using Kolmogorov's formula when $n = 3$, the function $\rho(a, b, c)$ of the coefficients could be written as composition of nine continuous functions of two coefficients each:

$$\rho(a, b, c) = g_1(y_1, c) + g_2(y_2, c) + \ldots + g_9(y_9, c)$$

where $y_q = \varphi_q(a, b)$. With a deft turn-around, Arnold refuted Hilbert's conjecture as a mere corollary!

Luck was not with Arnold and he was denied riding the tide of his discovery. Kolmogorov trumped this landmark result with his remarkable announcement in the same year, and all at once Problem 13 fell by the wayside. The break between Hilbert and Kolmogorov's unexpected result was now complete. He no longer even cited Hilbert's paper in the 1957 superposition *magnum opus* paper.

In conception and methodology, function representations embody a shift in paradigm. This raises the question: Why is Hilbert's Problem 13 mentioned in the title of a monograph whose centerpiece is a program that diverged from it? It is true that Problem

13 was stated as a specific problem in algebra aimed at computing roots of polynomial equations. But it did raise the broad question of the relation between the continuity of functions defined in Euclidean space and their number of variables, and in that it was ground-breaking.

That this should lead to computational algorithms and computer applications is one of those intriguing meanderings of a pregnant scientific idea. In addition, this is a fascinating story on its own merit. Hilbert had sowed the seed of the idea of dimension-reducing superpositions, and however weakly, the two stories are connected by an umbilical cord.

During the four decades following the publication of Kolmogorov's 1957 superposition theorem it remained mostly within the confines of mathematical theory. It was refined, extended to domains other than compact Euclidean spaces, but general computational algorithms failed to materialize, except for isolated sporadic efforts in the late 1980s and early 1990s. These did demonstrate the potential of function representations as an effective computational tool, but they did not generate a sustained momentum. These were important steps that broke the ice, so to speak, but systematic efforts to implement superpositions and derive computational algorithms did not follow. An impetus was needed to shift the thinking about function representations beyond their mathematical setting. Computer scientist Robert Hecht-Nielsen entered the story at that critical time and provided the needed impetus:

> At the First International Congress of Neural Networks held in San Diego, California, thirty years after the publication of Kolmogorov's result, Hecht-Nielsen moved Kolmogorov's function representations from mathematical existence into the realm of modern computing. He accomplished this by interpreting an alternative version of the formula as a feed-forward neural network (Hecht-Nielsen [40]).

He used the following equivalent version with functions composed of translates of a single continuous function $\psi(x)$ and powers of a suitable constant $\lambda$ (Sprecher [85]). In the case $n = 2$ this is:

$$y_0 = \psi(x_1) + \lambda\psi(x_2)$$

$$y_1 = \psi(x_1 + a) + \lambda\psi(x_2 + a) + 1$$

$$y_2 = \psi(x_1 + 2a) + \lambda\psi(x_2 + 2a) + 2$$

$$y_3 = \psi(x_1 + 3a) + \lambda\psi(x_2 + 3a) + 3$$

$$y_4 = \psi(x_1 + 4a) + \lambda\psi(x_2 + 4a) + 4$$

$$f(x_1, x_2) = g_0(y_0) + g_1(y_1) + g_2(y_2) + g_3(y_3) + g_4(y_4)$$

Hecht-Nielsen's network is shown schematically in Figure 6 for functions of two variables. Remember that continuity is an analogue concept, and so is Kolmogorov's superposition formula, but Hecht-Nielsen's interpretation suggested that it could be a basis for digital implementation, and applications did follow this time.

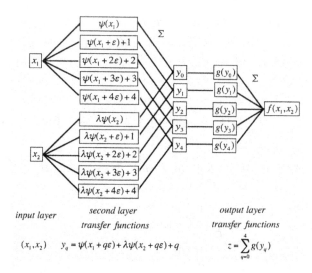

Figure 6: Hecht-Nielsen's Network when $n = 2$

His interpretation of Kolmogorov's function representations as a neural network shifted the paradigm again, from mathematical theory to implementation as a network. With the successful demonstration of the potential as a computation tool that was concurrent with it, two avenues were opened: adapting the formula to the specific requirements of neural networks, and deriving general computational algorithms.

Recognizing Kolmogorov's formula as the centerpiece of this narrative, it is featured ahead of Hilbert's Problem 13 that preceded it by half a century, because it provided the momentum for function-presentation based research. In itself, Problem 13 was in a vein that was detached both in its methodology and orientation from Kolmogorov's superpositions and the developments that followed.

Chapter Two presents initial attempts at function representations with the concept of mathematical trees. This gave the first clue to the possibility of representations of continuous functions in general with superpositions with fewer variables in general, and resulted in Arnold's refutation of Hilbert's conjecture. This is followed with a detailed anatomy of superpositions, their construction and their variations, and Chapter Three looks at aspects of function representations.

Chapter Four looks at superpositions as blueprints for Hecht-Nielsen's neural networks and other modes of computing, their implementation and applications. Chapter Five completes the story with a fuller discussion of Hilbert's Problem 13, including a second yet unresolved problem folded into its phrasing. Additional comments about this problem are deferred to Appendix B.

A common cliché in Euclidean geometry is that the shortest distance between two points is a straight line. There is no straight line that connects Hilbert's algebra problem of computing the roots of a polynomial equation to computational algorithms and modern computing.

The meandering path that we shall follow in reverse chronology allows us to approach the many aspects of the story from different perspectives. In some sense the chapters of this monograph constitute fairly independent essays, with a sufficient overlap to form a coherent narrative. Shifts in point of view of overlapping paragraphs were correlated to the developing narrative The order of chapters can be scrambled to some extent, and some sections can be omitted entirely.

There had been a renewed interest in Problem 13 in the wake of Kolmogorov's superposition theorem; a mention of it appears in many of the research articles and papers devoted to computing and application. Much of that renaissance and retrospective evaluation of Problem 13 views it through the kaleidoscope of superpositions. Some of these views tended to narrowly identify Problem 13 with the specific conjecture regarding the polynomial equation of the seventh degree. This often kept in the shadow or completely neglected the second and deeper problem embedded in Hilbert's problem; It concerns function representations under more stringent smoothness conditions than mere continuity, and it remains yet unsolved. The impact of research on this part of the problem on computing is unexplored to date. A brief account of this problem is part of our story.

A lively interest in all these topics continues, and new papers, articles and book chapters continue appearing in an increasing diversity of outlets that reflects spreading applications. Some sources are referred to in the manuscript for further reading and to indicate the diverse interests that the names Hilbert and Kolmogorov evoke.

A few mathematicians kept their eye on the kaleidoscope through several decades following Kolmogorov's discovery; they kept reframing their views through insights gained by focusing of different aspects of the problem. These evolving views are cited in the text, and the bibliography contains the appropriate references.

One wonders what turn the interest in Hilbert's Problem 13 would have taken if Kolmogorov's superpositions had not followed on the heels of Arnold's refutation. This author has no doubt that once Arnold opened the door to his version of representations with superpositions, a Kolmogorov -type discovery would have followed, perhaps even by Arnold who maintained an abiding interest in the problem.

Browsing the Internet reveals a rich source of information on the topics of this monograph. Google searches, lasting less than one minute each, yielded interesting information both about our time and about the level and proliferation of interests. A June 2016 count of web sites showed the following:

" Hilbert's 23 problems" 1,560 sites

" Hilbert's 13th Problem" 1,660 sites

" Kolmogorov superpositions" 1,720 sites

"Computational algorithms" 286,000 sites

The numbers were considerably larger in searches without quotation marks.

On a personal note, my own interest in superpositions began in 1959 as a graduate student of mathematics at the University of Maryland and continued through the intervening years. My interests were purely mathematical until a fortuitous pair of incidents changed the direction of my research to questions of implementation and computational algorithms.

This conversion began one day in 1993, when I stopped at the Earthling bookstore/café during a walk in downtown Santa Barbara to rummage through its shelves of timeworn paperbacks, some with cake crumbs inadvertently pressed between their pages.

One freestanding shelf was overflowing with an eclectic collection of paperbacks on topics such as the ABCs of computing, Fortran, Self-Improvement, Calculus for Idiots, MBA Made Easy and more My eye caught among them a gleaming new hardcover: *Neurocomputing* by Hecht-Nielsen, item [41] in the bibliography.

Intrigued by the title, I pulled it off the shelf and opened it to a random page. Staring in my face was my name and a section devoted to the version of Kolmogorov's formula of my University of Maryland dissertation-based paper appearing close over three decades earlier. It was then that I discovered its interpretation as a neural network, whatever that was. I bought the book to find out, and within two weeks I shook hands with Robert Hecht-Nielsen, a computer scientist at the University of California at San Diego. It turned out that the problem I was working on at the time would answer some of the questions on his mind, and this led to a still ongoing friendship with fruitful meetings that reoriented my research from mathematics to computer science.

Lightning struck again exactly one week after my first meeting with Hecht-Nielsen in the form of a telephone call from one Bryan Travis at the Los Alamos National Laboratory. He was trying for some time to write a computer program to implement the same formula that attracted Hecht-Nielsen's attention, and he needed help with some of the constructions; four weeks later I met Travis in Los Alamos. Also this was the beginning of a collaboration and lasting friendship. He was a pioneer in establishing the computability of function representations. Travis and Hecht-Nielsen did not know one another, and the juxtaposition of these two events was completely coincidental.

I recall a British algebra primer of around the year 1900 in which the author stated that his motivation for writing it was to free school children from having to make sense of the contrived problems in existing primers. His book would use only realistic problems, he stated in the preface, but the first problem in his book began: "*A dog is barking on the road to Nanking.*" I found that this was quite inevitable in a book describing mathematical theories and unintuitive constructs for which barking dogs on the road to Nanking may be the only example. I must concede their presence here and there in this story, but I tried my best to give the barks a musical ring.

And another *mea culpa*: In the introduction to the 23 problems, Hilbert quoted an old French saying:

> A mathematical theory is not to be considered complete until you have made it so clear that you can explain it to the first man whom you meet in the street. (Hilbert [58])

I accept without challenge that I failed to meet this goal, but in its spirit I have endeavored to make the narrative accessible to readers with only rudimentary knowledge of calculus, sacrificing for the purpose some technical detail, and often providing only a sketch of an idea behind a proof. The number of formal definitions and theorems has been has been kept low to achieve a more flowing narrative.

New York, 2016

# Chapter 1

# Historical Synopsis

The tantalizing story of the meandering of a problem in algebra to function representations and beyond to computer architecture and computational algorithms begins properly with the general seventh degree polynomial equation:

$$a_1 x^7 + a_2 x^6 + a_3 x^5 + a_4 x^4 + a_5 x^3 + a_6 x^2 + a_7 x + a_8 = 0$$

with arbitrary real number coefficients. Since different coefficients give different polynomial equations, they are regarded as parameters (variables), and with that meaning roots become functions of eight independent variable number-coefficients. Using only algebraic methods: the four arithmetic operations and radicals, this equation can be reduced to the form:

$$x^7 + ax^3 + bx^2 + cx + 1 = 0.$$

As explained in Chapter Five, the roots of this equation are functions of only three coefficients, and from these the roots of the original equations can be computed.

Hilbert's conjecture asserts that $m_\rho(3) = 3$ for the roots $\rho(a, b, c)$ of the last equation in the context of classification index $m_f(n)$ that was introduced in the Introduction. Note that

$m_\rho(8) = 3$ for the roots of the original equation. From this we could draw the general conclusion that for continuous functions of eight variables, $m_f(8) \geq 3$.

In fact, five coefficients can be eliminated by algebraic means from any general polynomial equation of degree five and higher, and this tells us that the number of coefficients (variables) of roots of general polynomial equations is not in itself a useful classification index.

Knowing this makes us wonder if the reduction of five coefficients in polynomial equations of degree five or higher is the best possible. Algebraic methods failed to do this in the case of the seventh degree polynomial equation, and after careful consideration Hilbert conjectured that also superpositions of continuous functions of two variables would fail. This is the conjecture that became known as Problem 13. As we mentioned in the preface, it was one of twenty-three problems that the rising star David Hilbert proposed at a lecture at the Congress of Mathematics held in Paris in the year 1900; it was also what led five decades later to Kolmogorov's representation theorem.

The chronology of this monograph should therefore have started with Hilbert's lecture, but instead we are leapfrogging fifty-six years ahead to begin with its outstanding and unexpected consequence. We did this to avoid being distracted by algebraic consideration whose language and methodology have no more than a ghostly contact with those of Kolmogorov's specific superpositions and representation theorem. Yet it is undeniable that this origin must be included in the complete story of function representations, especially because Hilbert combined continuous functions with the arithmetic operations and radicals that were the hallmarks of algebraic manipulations. This beginning, which makes for a fascinating story in its own right, is deferred to Chapter Five, where it provides an essential arc of the circle of ideas that led from algebra to modern computing and computational algorithms. What attracted Hilbert's attention to this particular equation and how he came to select it from among all others is part of the story that we will tell in Chapter Five. Here we merely mention the particular method for finding its roots that Hilbert introduced in the opening statement of Problem 13 and which enters a new term into our vocabulary: *Nomography*. Quoting from Hilbert's lecture as it appears in translation in (Hilbert [44]):

> Nomography[1] deals with the problem: to Solve equations by means of drawings of families of curves depending on an arbitrary parameter. It is seen at once that every of an equation whose coefficients depend upon only two parameters, that is every function of two independent variable, can be represented in manifold ways according to the principle lying at the foundation of nomography.

*Nomography* is an analogue graphical technique for finding solutions of equations using suitable parameters—not more than two at a time we are told. According to this, the roots of an equation having two coefficients can be computed with nomographic methods, as can every function of two variables, but the equation that Hilbert selected has three coefficients! To find its roots with nomography one of the coefficients had to be eliminated.

Some readers may recall other such analogue devices of yesteryear: the slide-rule and logarithmic paper are examples. For our purposes we don't have to delve further into nomography, but the interested reader may wish to consult Eversham's *Nomography* [32].[2]

The connection between Hilbert and nomography appears to have been through Paris. He first traveled to that city in 1886 under the urgings of the mathematicians Felix Klein and Adolf Hurwitz to

---

[1]d'Ocagne, Traité de Nomography, Paris, 1899.
[2]See also Epstein [31] and Ford [34].

broaden his horizons (Reid [81]). That city was at the time a lively hub of mathematics, and among the French mathematicians that he met during his visit was Maurice d'Ocagne, whose definitive book on nomography was published in 1891 (d'Ocagne [25]); Hilbert's probable introduction to nomography occurred at a time of growing interest in Germany and France in finding graphical methods for solving functional equations with fast computations and minimal errors.

The occasion of Hilbert's second trip to Paris was the International Congress of Mathematicians held there in the year 1900. Such congresses were inaugurated following an idea traceable to Felix Klein and Georg Cantor. Zurich was the site of the first congress held in in 1897; it is a historical coincidence that this was also the year of the first Zionist Congress convened by Theodor Herzl in Basel. The second congress was held in Paris in 1900, and since then congresses were held regularly every four yeas in different cities, with the exception of the two World Wars, when none were held.

David Hilbert, born in 1862, was quick to establish the reputation of an emerging mathematician of note. He developed early disdain for intellectual narrowness, and when he was invited to deliver a lecture at that congress he selected an audacious program featuring 23 problems that he heralded in the opening program of its 40 or so pages simply titled *Mathematical Problems:*

> Who of us would not be glad to lift the veil behind which the future lies hidden; to cast a glance at the next advances our science and at the secrets of the development during future centuries? ...We know that every age has its own problems which the following age either solves or casts aside as profitless and replaces with new ones (Hilbert [44]).

These problems, shrouded with this lofty mantle showed Hilbert's formidable knowledge of the mathematics of his day and a keen sense of where its inventive energy should be directed He considered the problems that he proposed essential for mathematics to solve during the coming century. Hilbert consulted on this idea with Hermann Minkowski, the same Minkowski whose lectures Albert Einstein failed to attend when he studied at the Swiss Federal Polytechnic in Zurich (later the ETH), and who changed our perception of space and time forever when he formulated Einstein's Special Theory of Relativity in four-dimensional space.

Hilbert was slow in producing his rather long lecture, and only 10 problems were available by the time of his talk; Problem 13, which is our focus, was included among them as Problem 6. Eventually all twenty-three problems saw the light of day. An excellent account of the circumstances of Hilbert's lecture can be found in (Grattan-Guiness [38]).

As of this writing, four of the twenty-three problems remain unsolved; some problems had been partially solved, and some solutions are controversial. Problem 13 is among those listed as partially solved for the reason explained later. Other than describing this problem in some detail, this manuscript is devoted to the unexpected developments that it brought about.

Hilbert presented what appears on the surface to be a mundane problem to an august audience of mathematicians in a formal setting; this suggests caution in jumping to conclusions, especially in view of this excerpt from Hilbert's concluding remarks to his lecture:

> The problems mentioned are merely samples of problems, yet they will suffice to show how rich, how manifold and how extensive the mathematical science of today is. . .

So also this problem was intended to move mathematics forward, to open window for further discoveries. It was the seed, so to speak, of unintended and unimagined developments that followed over five decades later—a short time-span in a chronology of science but a long time on a human calendar.

Before leaving this capsule summary we summarize what we have learned, beginning with the important observation that Hilbert selected the roots of the normalized seventh degree polynomial equation because he saw in them examples of algebraic functions of three variables not representable with superpositions of continuous functions of two variables. That is, he perceived in these to be functions of three variables that cannot be further reduced through superpositions and can therefore be thought of as functions of three variables, implying that $m_f(3) = 3$ as explained in the preface.

Hilbert's conjecture raised obliquely a broader question encompassing the connection between dimensionality and functional complexity that took center stage in a different form five decades later. Even though Problem 13 was aimed at a classification of algebraic functions based on variable-reducing superpositions, the classification of continuous functions more generally was lurking in the background, and this is where Kolmogorov and Arnold entered the narrative.

# Chapter 2

# Kolmogorov's Function Representations

## 2.1 Introduction

In the story of function representations, algebra and mathematical analysis shared axiomatically and methodologically a tentative contact point at which algebra opened a window for a new insight in analysis. It was actually not algebra per se that set these discoveries in motion, but Hilbert's shrewd insight and choice of superpositions as a methodology for making an algebraic point.

We know what that point was: That the computations of algebraic functions (roots of polynomial equations) of three variables cannot be broken down with the tools of algebra into steps involving only two variables at a time; that superpositions of continuous functions as the alternative methodology would also fail to accomplish this. Therefore, there exist continuous functions $f$ such that $m_f(3) = 3$.

By extension this can be viewed as a surrogate for the general problem of breaking down computations of continuous functions with many variables into finitely many steps involving fewer variables. Whereas superpositions were for Hilbert only a tool, Kolmogorov made the resulting concept of function representations into the object of a general study of continuous function.

Once he shifted the point of view, the solution of Hilbert's Problem 13 became only one aspect of the study of function representations. As we already mentioned, algebra, which was Hilbert's concern, fell by the wayside when function representations took an independent life of their own, after Kolmogorov's famous representation formula delegated the original impetus for the study of superpositions to a footnote.

As I review these comments I am left with the uneasy feeling that the narrative resembles a cat chasing its own tail, and that the concepts of superpositions and function representations require greater clarity before exploring them in greater detail and depth. For one thing, the distinction between these two concepts must be clarified: A superposition is the methodology of composing functions of functions that are the building blocks for equivalent representation of continuous functions.

For the second time I take the risk of being overly pedantic with a brief recall of a lesson from calculus: real-valued continuous functions defined on a Euclidean $n$-dimensional unit cube $E^n$ are commonly described symbolically: $f : E^n \to \mathbb{R}$, where $\mathbb{R}$ represents the real line. Such a function associates a unique real number $y \in R$ with every point $(x_1, \ldots, x_n) \in E^n$. A common feature of function representation is making this association an indirect two-step operation as shown in Figure 2.1.

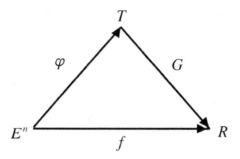

Figure 2.1: Commuting Diagram of the Function
Representation $f = (G(\varphi))$

The direct route $f : E^n \to \mathbb{R}$ constitutes the base of the diagram. In the first step the original spatial variables $(x_1, \ldots, x_n)$ are replaced by new variables $(t_1, .., t_m)$ through an intermediary mapping $\varphi : E^n \to T^m$

$$\varphi : (x_1, .., x_n) \to (t_1, .., t_m)$$

This step is determined by the chosen strategy. It relates points in the $n$-dimensional Euclidean space $E^n$ to points in an $m$-dimensional Euclidean space $T^m$. This step is completely independent of $f$.

In the second step, the target function $f$ is computed with the new variables

$$t_1, .., t_m : G : T^m \to \mathbb{R}$$

yielding

$$G : (t_1, \ldots, t_m) \to f(x_1, \ldots, x_n)$$

The resulting outcome is schematically represented as $G(\varphi) : E^n \to \mathbb{R}$ in Figure 2.1.

## 2.2   Enter Kolmogorov

Two pivotal results in the preface were Kolmogorov's *opus magnum* representation of functions, and Arnold's refutation of Hilbert's conjecture, both appearing in 1957 in reverse order. As already mentioned, the two papers in which these results appeared were preceded by a seminal 1956 paper of Kolmogorov, and the three papers form an organic trio. That paper was aimed at finding a lower bound on variable-reducing representations of continuous functions of several variables, but it failed to answer Hilbert's conjecture.

Arnold's paper that was based on it had a limited scope. Rather than investigating function representations at large, as Kolmogorov did, it was focused on Hilbert's conjecture. He refuted it by proving that $m_f(3) = 2$ for all continuous functions of three variables, making the algebraic setting of the conjecture irrelevant.

A change in paradigm sets the Kolmogorov's 1957 paper apart from these two because of its complete break with Hilbert's conjecture: Although this paper grew out of the former two, it did not share their aims or methodology. Still, some authors, notably Vituškin, see this as a collaborative effort, and refer to KRT as Arnold - Kolmogorov Representations Theorem.

To prevent erroneous conclusions, it must be emphasized that all of these considerations are limited to continuous functions defined on the Euclidean unit cube $E^n$.

These three papers were short announcements with incomplete constructions and proofs, as is not unusual with papers appearing in *Doklady Akademii Nauk SSSR*. Their sparseness of cited references is noteworthy: Other than citing their own papers and Hilbert's problem that triggered it all, only two other works were cited.

Kolmogorov's four-page 1956 paper was titled *On the Representation of Continuous Functions of Several Variables by Superpositions of Continuous Functions of a Smaller Number of Variables* (Kolmogorov [54]).

This, like the title of Kolmogorov's other paper, is explicit in describing its content. It cited three references of which only Kronrod's was relatively recent; the first was Hilbert's 1900 lecture in his collected works, and the second was on a subject not previously connected to this circle of ideas. The references were as follows:

1. David Hilbert: *Gesamelted Abhandlungen* (3)17, Springer, Berlin 1935.

2. Karl Menger: *Kurventheorie*, Teubner, Leipzig-Berlin 1932.

3. Alexander Semenovich Kronrod: *On functions of two variables*, Uspehi Mat. Nauk 5(1)35, 1950;

Arnold's three and a half page 1957 paper bore the brief title *On functions of three variables;* it cited two references: Kolmogorov's 1956 paper and Menger's paper. Kolmogorov's four and a half page 1957 paper cited only two references: his 1956 paper and Arnold's paper.

This sequence of three short announcements appeared in breathtaking staccato rhythm as seen from a comparison of the issues of *Doklady* in which they appeared one year apart: Kolmogorov's first paper appeared in issue 108, and both Arnold's and Kolmogorov's second paper appeared in issue 114.

The cited Menger's 1932 *Theory of curves* and Kronrod's 1950 paper appeared more than three decades after the statement of Hilbert's Problem 13. A clue of interest in the general area of superpositions can be seen in the already mentioned the Nina Bari 1930 paper: *Memoire sur la representation finie des fonctions continues*, that appeared in the Mathematische Annalen (Bari [8]); Tschebotarow's 1954 paper *Über ein algebraisches Problem von*

*Herrn Hilbert* (Tschebotarow [13]); and Vituškin's 1954 paper *On Hilbert's 13th problem* (Vituškin [105]).

Kolmogorov published in 1955 a paper on representability of functions of several variables by superpositions of functions of fewer variables but the trio of the 1956-57 papers does not refer to it. Judged by the only three source works that they referenced, these papers were self-contained and removed from other on-going research as far as we know. Arnold, in a position to know about the history of Kolmogorov's interest in Hilbert's Problem 13, said only that Kolmogorov was working on it (Arnold [5]). But he offered invaluable insight on Kolmogorov's thinking and way of adopting a set-theoretic approach to the structure of continuous functions using trees. Vituškin, another student of Kolmogorov, published in 1954 a paper on Hilbert's Problem 13. This and a seminar indicate that Kolmogorov's interest in it dates back at least that far.

Arnold's nine-summand formula that disproved Hilbert's conjecture was featured in the preface It may have appeared bewildering! Its geometric interpretation is explained below, and for now we observe that it is a mathematical existence theorem: A proposition that tells us that a certain decomposition of functions into superpositions is possible under the stated conditions, but it does not give us a clue about how to actually find the functions $\varphi_q$, and the functions $g_q$ when a specific function $f$ is given. Just like knowing that there is gold in them thar hills does not help us find it.

His result went beyond refuting Hilbert's conjecture by proving a more general fact about continuous functions of three variables. We repeat an earlier enjoiner that no examples of a representation of a concrete function of three variables can be provided because the proof of the theorem establishing the formula is not constructive, and the formula is not computational.

This brings us to an irony in Arnold's refutation of Hilbert's conjecture: It stemmed from an attempt to compute the roots of the polynomial equation $x^7 + ax^3 + bx^2 + cx + 1 = 0$ by analogue graphical means, leading Hilbert to conjecture that *in theory this could not be done*. Of course Hilbert's interest was purely theoretical and actual computation was not at issue. Arnold showed in contrast that *in theory it could be done*, but in the way he did it, the issue of computation that led to the problem fell by the wayside. It remained unresolved in Arnold's paper in the sense that no algorithm that would actually compute these roots was part of his solution.

It is most important to understand that the concerns of both Hilbert and Arnold (and Kolmogorov ) were mathematical theory and not computability. Theory establishes certain logical conclusions, but the development of computational algorithms and practical usage do not always follow, and require other techniques and considerations.

The surprising announcement that Hilbert's conjecture was wrong received no more of a notice than a footnote because of the juxtaposition of Arnold's and Kolmogorov's results with a shift in paradigm in the same issue of the Doklady, as we already noted. Kolmogorov's theorem included a slightly improved version of Arnold's formula described below, but it went far beyond his findings by showing that that there were no continuous functions of more than one variable, in the sense they were compositions of continuous functions of one variable and addition. I hasten to add that unless otherwise indicated, the functions we refer to are real-valued continuous functions defined on a compact (closed and bounded) domain, usually the unit $n$-cube $E^n$.

In a different language, Kolmogorov's result said that the concept of functional complexity, as far as continuous functions were concerned, was dimension-independent when the measure of complexity was finitely many superpositions of continuous functions; that is, $m_f(n) = 1$. In the context of all continuous functions with domain $E^n$ this runs counter to intuition.

Indeed, former students of calculus may recall that the study of functions of two or more variables was technically but not conceptually more difficult than that of functions of one variable. Many concepts such as derivatives and integrals, generalized in a natural way, except that derivatives were replaced by partial derivatives and directional derivatives.

Functions of more variables were often analyzed by looking at cross-sections or holding all but one of the variables constant. It appeared intuitive that some of the technical difficulties when dealing with functions of more variables were inherent in the number of variables; it seemed that functional complexity was tied to this in some way.

Kolmogorov's was an astonishing discovery, and combined with his eminence it was not surprising that it shunted Hilbert's Problem 13 and Arnold's refutation to a sidetrack. Despite the disparity between the status of Arnold and Kolmogorov, however, Arnold's paper, and independently Hilbert, might have merited more attention then they did. Disposing of one of Hilbert's problems fifty-seven years later was still a feat, especially when Hilbert was proved to be wrong!

Kolmogorov's function presentations refuted Hilbert's conjecture as a mere corollary, and so also was Arnold's result. Kolmogorov did not connect his result directly with this conjecture, and as we already noted there was no mention of that anywhere in the paper. He did add the following note immediately after the statement of his theorem:[1]

When n=3 we set

$$\varphi_q(x_1, x_2) = \psi_{1,q}(x_1) + \psi_{2,q}(x_2),$$

$$h_q(y, x_3) = g_q[y + \psi_{3,q}(x_3)]$$

[1] Kolmogorov [55]

then [ Arnold's] formula can be written in the form

$$f(x_1, x_2, x_3) = \sum_{q=1}^{7} h_q[\varphi_q(x_1, x_2), x_3],$$

which is a slight strengthening of the result of V. I. Arnold.

Significantly, he did not connect Arnol'd formula with Hilbert's problem.

The reviews that these two papers received in the *Mathematical Reviews*, the organ of the American Mathematical Society that publishes peer reviews of published research around the world, was consistent with the tradition of focused, factual analysis. Three extraordinary points might have been highlighted in the review of Arnold's' paper, though: That all continuous functions of three variables are superpositions of continuous functions of two variables; that another one of Hilbert's problems had been solved; and that the solution was negative, but none were. Instead, the paper received a seven-line review:

Brief exposition of proof that any continuous real function $f(x_1, x_2, x_3)$ defined on the unit 3-cube can be represented in the form:

$$f(x_1, x_2, x_3) = \sum_{i=1}^{3} \sum_{j=1}^{3} h_{ij}[\varphi_{ij}(x_1, x_2), x_3]$$

where $h_{ij}$ and $\varphi_{ij}$ are continuous real functions of two variables.

This result solves the famous "13th problem of Hilbert ", in the sense of refuting the conjecture there stated. The corresponding result for functions of more than 3 variables was obtained by A. N. Kolmogorov. [MR0111808 (22, 2668)]

This 'for the record' announcement made no note of this triple accomplishment. Neither was there a description of Arnold's methodology, which was novel enough to be mentioned in view of the algebraic underpinning of Hilbert's problem. The same reviewer devoted 23 lines with considerable detail to the review of Kolmogorov's 1957 paper, including the specifics of the strengthening of the result of Arnold noted earlier, and again without connecting this with Hilbert's Problem 13 [MR0111809 (22, 2669)].

Kolmogorov's 1956 paper had also received a detailed review of corresponding length, but the unexpected conclusion, that there were no continuous functions of more than three variables, was mentioned at the end in the nature of a footnote [MR0080129 (18, 197b)].

Recall that Problem 13 was not concerned with function representations or the structure of continuous functions per se. It was concerned with the structure of algebraic functions insofar as they could be built with simpler components. Kolmogorov's 1956 paper did contain the unexpected result that every continuous function of four or more variables was representable with superpositions of continuous functions of three variables, but this lower bound did not allow him to touch Hilbert's conjecture.

In an important way, these responses reflect the state of mathematics of the time. This was a period of unprecedented expansion of mathematics, in many countries and especially in the United States, with research subsidized by the armed forces and the National Science Foundation. What had fallen by the wayside was the true context of Hilbert's Problem 13, and that its analytic part remained open. That this conclusion extended more generally to Problem 13 is evident from the sparse attention it received in earlier years.

The interest that Kolmogorov's theorem evoked in the mathematical community can be gauged by the more than one hundred and fifty publications between 1955 and 1987. As far as implementing superpositions and applying them, however, Kolmogorov's theorem remained an existence theorem with few exceptions as late as 1987. Research that immediately followed its publication was aimed at the different aspects of superpositions: proofs and alternative formulations; the anatomy of superpositions and inner functions; extensions to unbounded functions and other spaces; number of summands.

George. G. Lorentz was the first to publish an elegant proof (Lorentz [67]) that smoothed some of the rough edges in Kolmogorov's outlined proof, though not entirely. Hedberg also provided a notable proof in a broader setting (Hedberg [42]) and Kahane provided the simplest but most abstract proof (Kahane [65]) An innovative proofs of Fridman [35] developed an alternative construction that improved the inner functions. Ostrand generalized the superposition theorem from $E^n$ (and other compact spaces more generally) to metric spaces (Ostrand [76]); Doss extended it to unbounded functions (Doss [30]) And there was other research, notably by Vituškin and Henkin, aimed as superpositions with smooth (differentiable) functions.

## 2.3 The Kolmogorov-Arnold Papers

We saw in the introduction how Kolmogorov approached Hilbert's conjecture indirectly through a general method of representation of continuous functions of any number of variables. His strategy for solving the conjecture breached the dyke of dimensionality, leading to function representations with superpositions of functions of three variables, but excluded superpositions with functions of two variables.

After making the breakthrough, Kolmogorov passed the problem to Arnold who brought Kolmogorov's partial success to fruition by showing the all continuous functions of three variables are representable with superpositions of continuous functions of two variables; that is $m_f(3) = 2$ and giving thereby a direct refutation of Hilbert's conjecture.

Analyzing his own and Arnold's work, Kolmogorov was quick to realize that this was not the end of the story, and he bested both Arnold and Hilbert by shifting the paradigm: The conjecture was no longer of direct interest and superpositions in their role in representation of continuous functions of any number of variables $n \geqslant 2$ took an independent life of their own.

This capsule summary outlines the fascinating story leading up to Kolmogorov's 1957 function representations In a significant way, Kolmogorov took the idea of superpositions to its ultimate formulation in a format and elegance that defied further refinement.

This result represented a complete break with Hilbert's thoughts as he articulated them in Problem 13, and this point of view explains Kolmogorov's omitting mentioning of his name in his paper. The break from the formulation of that problem in setting an independent agenda is absolute in the sense that also the methodology leading up to Kolmogorov's function representations was a break from classical analysis that was the backdrop of this problem. He accomplished the end result geometrically, with astonishingly elementary means.

The first important discovery in the genealogy of dimension-reducing function representations was the third theorem in Kolmogorov's 1956 paper. This breakthrough was based on Karl Menger's *universal tree* containing a copy of every other tree, and on the discovery of Kronrod connecting Menger's trees with level curves of continuous functions. Tracing these steps takes us now to terra incognita; on a journey through the strange landscape of mathematical trees.

A simplistic mental image was depicted in Figure 2.5;[2] a tree denuded of leaves as a child might draw with short strokes of line segments whose one end emanates from a branch and whose other

---

[2]See also Arnold [6].

end is free and does not touch any part of the tree. In simplest terms, a tree is a locally connected continuum not containing a homeomorphic image (copy) of a circle; that is, a tree has no intersecting branches.

Before pursuing this further, we should sidestep to understand the emergence of the concept of mathematical tree, and this may be best explained in the context of Hilbert's algebraic setting.

The goal of reducing the number of coefficients of polynomial equations was to simplify the finding of roots. This was accomplished by manipulating the polynomial s through algebraic operations; that is, transforming their structure to similar structures. From an analytic point of view, the transformation from

$$P_1(a_0, a_1, ..., a_7) =$$
$$a_0 x^7 + a_1 x^6 + a_2 x^5 + a_3 x^4 + a_4 x^3 + a_5 x^2 + a_6 x + a_7$$

to

$$P_2(b_1, b_2, b_3) = x^7 + b_1 x^3 + b_2 x^2 + b_3 x + 1$$

might be regarded as one of form rather than substance. Dimension considerations played no role here. In other words, instead of being a goal, function representations in this context were the means for simplifying the computation of roots.

The significant difference between this and the transformation from $f(x_1, ..., x_n)$ to

$$\sum_{q=1}^{2n+1} g_q \circ \sum_{p=1}^{n} \psi_{pq}(x_p)$$

is that it is an end in itself; using these for computation was not a goal.

Here a different strategy is called for. This entails a transformation of the domain of definition, the unit cube $E^n = E \times E \times ... \times E$ in our case; that is, finding continuous transformations $\varphi : E^n \to E^m$ with $m < n$ for constructing functions $g(\varphi)$ that compute our $f$. This implies, of course, that these transformations are independent of the functions to be represented. This also means that the $n$-dimensional cube must be arranged and 'packed' in a certain way into a lower dimensional cube, or the interval $E$ in the case of Kolmogorov's final representation formula. It is this packing that we will explain in the sequel that is the key to dimension-reducing function representations, and the subject of Kolmogorov's 1956 paper.

The simple Y tree shown in Figure 2.2 will serve to explain a little the application of this concept and the attraction of the method in the quest of dimension reduction. This tree consists of three segments that meet at a single intersection point. None of these line segments is distinguished as a trunk, and this term is not part of our tree vocabulary.

Let your imagination roam and think of a tree that has any number of branches emanating from any number of intersections. Every branch is one-dimensional and the concept of a tree broadens our idea of one-dimensionality and one-dimensional configurations.

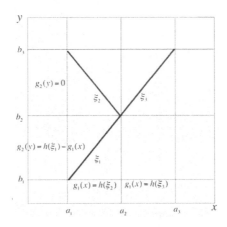

Figure 2.2: $h(\xi) = g_1(x) + g_2(y)$

To push this idea a little further, imagine a function defined on this tree. Its values would be traces along line segments. If the branches crowd against each other they would tend to 'linearize' the plane, and a function defined on it would be evaluated along its one-dimensional branches. We illustrate this idea with the Y-tree in Figure 2.2.

The values of a function $h(\xi)$ on the Y tree are specified separately on each branch by means of functions of the Cartesian coordinates $x$ and $y$ defined piecewise in the plane. For simplicity and clarity we label the variable $\xi$ differently on each branch: $\xi_1$, $\xi_2$, and $\xi_3$ as indicated in Figure 2.2.

Define the functions

$$g_1(x) = \begin{cases} h(\xi_2) & a_1 \leqslant x \leqslant a_2 \\ h(\xi_3) & a_2 \leqslant x \leqslant a_3 \end{cases}$$

$$g_2(x) = \begin{cases} h(\xi_1) - g_1(x)b_1 \leqslant x \leqslant b_2 \\ 0 & b_2 \leqslant x \leqslant b_3 \end{cases}$$

Note that Cartesian coordinates are used to designate points $\xi$ on the tree. The verification that this specification gives $h(\xi) = g_1(x) + g_2(y)$ in the tree topology is derived directly from the definitions of $g_1(x)$ and $g_2(y)$. Presented here is the most rudimentary idea about the subject that becomes, of course, complicated with complex trees and applications. Other examples will be presented below.

For now, just remember that a point $\xi$ on the tree in Figure 2.2 can be identified with coordinates $(g_1(x), g_2(y))$. The attraction to this method is, of course, the natural presentation $h(\xi) = g_1(x) + g_2(y)$ for points on a tree. This function can be viewed in two ways: As a function of the variable $\xi$ that varies over a Y- tree in the plane, and as a function specified by Euclidean coordinates.

The important concept to be taken away from this example is the possibility of a function that is defined piecewise on line segments of a tree that can be made as dense as we please and which, in turn, can be described, branch by branch, by means of functions specified with Cartesian coordinates

We need to know more about trees and how they figure in describing the graph of a continuous function of two or more variables. We explain this in the two dimensional case, introducing these new methods and way of thinking through the making of a commonplace topographic map. A segment of a landscape in upper New York State is depicted in Figure 2.3. The contours outline shapes and also their elevation above sea level.

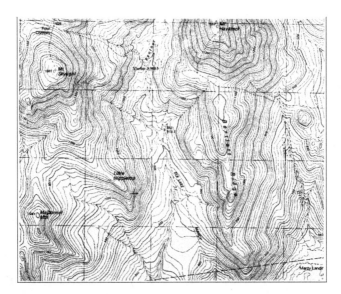

Figure 2.3: Topographic Map of a Mountainous Landscape

Think of a simpler landscape featuring two mountains, one with a double peak (Figure 2.4). The contour lines indicate an imaginary stack of evenly spaced horizontal curves of equal height above the primary plane and this introduces a new term into our vocabulary: *level sets*. This term refers to sets of points at which the graph of the function has the same constant value.

Figure 2.5 shows a bird's eye view of the projections of these level sets onto the plane, giving the essence of a topographic map of this landscape. Observe that in this figure some level sets form a continuous loop, others consist of two loops, and in two cases loops gave a contact point. The approximate shape of the landscape can be reconstructed from these projections by pulling the contours up the appropriate distance and covering them with a film.

The elevation above the plane of the contour lines up to the split into two peaks can be marked on a line segment that replaces the customary vertical axis. The elevation of each of the peaks can similarly be marked on two line segments that form a Y as in Figure 2.6.

Note that knowing a level set and its elevation above the plane is not enough to locate a specific point on the landscape: this requires of course three coordinates: $(x_1, x_2, y)$. Remarkably, however, we are able to represent a graph of a two-dimensional surface in three-dimensional space by means of level sets in a plane, and a one-dimensional graph, namely a tree. Indeed, Kronrod has demonstrated the connection between level sets and trees.

It is easy to visualize that the accuracy of the reconstructed landscape increases with increasing density of the level sets. As the distances between level sets diminish to zero, they fill the square, with a resulting tree that is more complex. On this tree a function $f(x_1, x_2)$ is represented through the level sets of its graph, and these heuristic considerations will help to understand Kolmogorov's 1956 paper that we are going to look at next. This is where the genealogy of Kolmogorov's superposition formula begins with his first function representation formula that we introduce in the case of two variables.

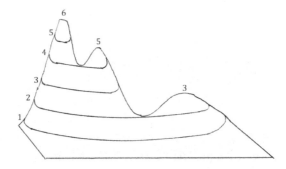

Figure 2.4: Landscape with Level Sets

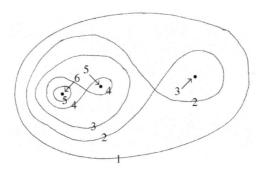

Figure 2.5: Projection of Level Sets into a Plane

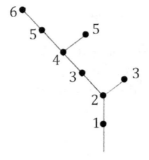

Figure 2.6: Tree Showing the Level Sets

Despite the incompleteness of the information as far as locating points is concerned, these considerations give a good overview and insight into alternative ways of looking at data in three-dimensional space through dimension reduction.

The application of tree appears nonintuitive on first encounter, but looking at the above graphs this begins to make sense. There is a considerable distance between these basic notions and their translation into rigorous mathematical formulation and actual quantitative tools, but we don't have to travel down that road since all we need here is qualitative understanding.

We now have two new concepts: one of dimension reduction through level sets, and the other of 'linearization' along branches of a tree. Each of these gives partial information about the graph of a function (or landscape in the preceding discussion). Both diverge conceptually from traditional geometric thinking of many readers along lines $y = f(x_1, x_2)$ in which height at a point $(x_1, x_2)$ is measured along the vertical y-axis. This gets us into the mathematical area of topology that is concerned with the study of geometric properties of spatial relations.

The discussion of level sets and trees involved the two dimensional plane containing the projections of the level sets, and instead of the usual vertical axis that measures height we have a more complicated one-dimensional configuration that gives in addition to height also partial location information. The landscape (graph) is in the three-dimensional cube $E^3$ that we would normally think of as $E^3 = E^2 \times E$, where location is determined at a point $(x_1, x_2) \in E^2$, and elevation is measured at a point $y \in E$. In our example the landscape was described in the cross-product space $E^2 \times \Xi$ in which the traditional complete description of a point on the graph is $(x_1, x_2, y)$ is replaced by the partial description $((x_1, x_2), \xi)$.

With these examples we are ready to examine the first of two findings leading to the *"unexpected consequence"* in Kolmogorov's

1956 paper He formulated the general case for arbitrary $n$ and for simplicity we revert to the case $n = 2$. The result in question was the discovery of three continuous functions:

$$\xi_1 = g_1(x_1, x_2)$$
$$\xi_2 = g_2(x_1, x_2) \tag{2.1}$$
$$\xi_3 = g_3(x_1, x_2)$$

with values in the universal tree $\Xi$. With these, every real-valued continuous function $f(x_1, x_2)$ defined on the unit square $E^2$ has a representation

$$f(x_1, x_2) = h_1(\xi_1) + h_2(\xi_2) + h_3(\xi_3) \tag{2.2}$$

with three continuous real-valued functions $h_q(\xi_q)$. We point out the obvious again: that the functions (2.1) are independent of $f$, and can be used to represent any continuous function $f$ defined on $E^2$.

The twist in this representation is that the functions $g_q(x_1, x_2)$ give the Cartesian coordinates of points $(x_1, x_2)$ on the tree $\Xi$; they are defined on the unit square, but their values are in $\Xi$. A hidden feature of this situation is that the functions $h_q(\xi_q)$ are continuous on $\Xi$ and the functions $h_q[g_q(x_1, x_2)]$ are continuous on the square $E^2$. The functions $h_q$ must therefore be continuous in both topologies.

Think about these functions in terms of level sets as in the above example. As we pointed out there, we surmise that a single function $h_q(\xi_q)$ cannot completely describe $f$ at every point $(x_1, x_2)$ in its domain, but this is made possible by the juxtaposition of three tree configurations.

What do the representations in equations (2.1) and (2.2) accomplish? First and foremost the functions $g_q(x_1, x_2)$ are independent of the target function $f$. As was noted earlier, this is the hallmark of function representations. Second, even though these functions have the same number of variables as $f$, the functions $h_q(\xi_q)$ that compute it are functions of one variable only. This is an example of the 'packing' that we mentioned earlier. The two-dimensional square is 'packed' into the one-dimensional tree.

Saying that the functions $h_q(\xi)$ compute $f$ is of course only a manner of speaking, because like the other theorems that have been mentioned so far, also this is an existence theorem: it tells us that formulas (2.1) and (2.2) are valid, but not how to actually use then to compute a function $f$. The formula is represented schematically with the commuting diagram in Figure 2.7:[3]

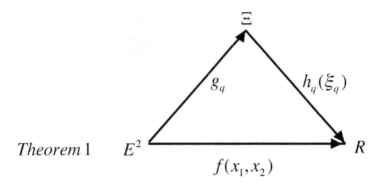

Figure 2.7: Commuting Diagram for Representation (2.1)–(2.2)

Recalling Figure 2.2 depicting a Y tree, we can think of the universal tree $\Xi$ as an intermediate agent in the sense that the transformation $g_q : E^2 \to \Xi$ arranges the (unordered) points $(x_1, x_2)$ of the square $E^2$ in a way that enables the functions $h_q$ to compute $f$ on the one-dimensional tree. Note that the input values $(x_1, x_2)$ and output values $f(x_1, x_2)$ are real numbers. The system of equations (2.1) being over determined, there is a relationship between $\xi_1$, $\xi_2$ and $\xi_3$, meaning that there is a certain coordination between them as a point $(x_1, x_2)$ traces a path through $E^2$.

---

[3]See also Figure 2.1.

In its full generality, Kolmogorov proved:

**Theorem 1.** *For each integer $n \geqslant 2$ there exist $n + 1$ continuous functions*

$$\xi_1 = g_1(x_1, ..., x_n)$$

$$\vdots \qquad\qquad (2.3)$$

$$\xi_{n+1} = g_{n+1}(x_1, ..., x_n)$$

*defined on $E^n$ and with values in the universal tree $\Xi$, such that every real-valued continuous function $f$ defined on $E^n$ has a representation*

$$f(x_1, ..., x_n) = \sum_{q=1}^{n+1} h_q(\xi_q) \qquad (2.4)$$

*with continuous functions $h_q$ on $\Xi$.*

The functions $h_q$ have a dual existence: They are continuous on $E^n$ and on $\Xi$. In this scheme, functions $g_q(x_1, ..., x_n)$ map the cube $E^n$ onto the universal tree $\Xi$. Looking at $E^n$ as the Cartesian product $E^n = E^{n-1} \times E$, Kolmogorov now argued that if the cube $E^{n-1}$ would be mapped onto $\Xi$ with functions $g_q(x_1, ..., x_{n-1})$, then $x_n$ could vary independently over the interval and a representation could be obtained in the product space $\Xi \times E$ instead of $\Xi$. He thus obtained the representation:

**Theorem 2.** *For each integer $n \geq 3$ there exist $n$ continuous functions*

$$\xi_1 = g_1(x_1, ..., x_{n-1})$$

$$\vdots \qquad\qquad (2.5)$$

$$\xi_n = g_n(x_1, ..., x_{n-1})$$

*defined on $E^{n-1}$ and with values in the universal tree $\Xi$, such that every real-valued continuous function $f$ defined on $E^n$ has a representation*

$$f(x_1, ..., x_n) = \sum_{q=1}^{n} h_q(\xi_q, x_n) \tag{2.6}$$

*with functions $h_q(\xi_q, x_n)$ that are continuous on the product $\Xi \times E$*

One variable now varies over an interval of real numbers, but there was a price to pay: Whereas the original representation applied to functions of $n \geq 2$ variables, the change of the basic space from $\Xi$ to $\Xi \times E$ required the lower bound $n \geqslant 3$.

This representation is the subject of the second discovery of the 1956 paper. It succeeded in reducing the number of variables in superpositions from $n$ to $n - 1$, but this gain continued the intermediate dependence on the universal tree $\Xi$.

For simplicity and easier comparison with the formulas (2.1) and (2.2), let's examine this representation for the lowest value $n = 3$:

$$\begin{aligned} \xi_1 &= g_1(x_1, x_2) \\ \xi_2 &= g_2(x_1, x_2) \\ \xi_3 &= g_3(x_1, x_2) \end{aligned} \tag{2.7}$$

$$f(x_1, x_2, x_3) = h_1(\xi_1, x_3) + h_2(\xi_2, x_3) + h_3(\xi_3, x_3) \tag{2.8}$$

This formula is explained with a commuting diagram in Figure 2.8. Our immediate observation is that the functions $g_1$, $g_2$ and $g_3$ that were functions of two variables in (2.1) are still functions of two variables, but now they compute a function of three variables. Also these functions have values in the universal tree $\Xi$;

that is, $g_q : E^2 \to \Xi$, and the functions $h_q(\xi_q, x_3)$ are real-valued and defined and continuous on the product space $\Xi \times E$.

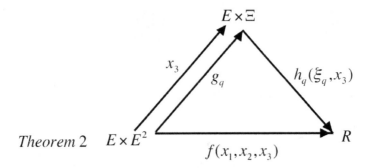

Figure 2.8: Commuting Diagram for Representation (2.7)–(2.8)

We now arrive at the first important breakthrough: it is a consequence of the third theorem in this 1956 paper. Falling short of settling Hilbert conjecture by the width of a hairbreadth, it had a far reaching and seminal outcome. In Kolmogorov's words:

> From Theorem 2 ... there follows a rather unexpected consequence: every continuous function of an arbitrarily large number of variables can be represented in the form of superpositions of continuous functions of not more than three variables.

Kolmogorov was now able to dispense with the universal tree and obtain a representation completely within the realm of real-valued continuous functions. There was again a price to pay: the lower bound that increased from $n \geqslant 2$ in the first representation, and then to $n \geq 3$ in the second, now increased to $n \geqslant 4$; that is, these representations applied only to functions $f(x_1, \ldots, x_n)$ of four or more variables.

To understand what follows, remember the Y tree and the natural presentation $h(\xi) = g_1(x) + g_1(y)$ of points on a tree. In particular, review how Euclidean coordinates specify the variable

$\xi$ that varies over a Y-tree in the plane. This is coupled with Kolmogorov's observation that the universal tree $\Xi$ can be realized as a tree in the unit square.

We introduce Theorem 3 in the lowest meaningful case, $n = 4$. The representations are based now on four sets of two continuous functions:

$$
\begin{aligned}
y_{1,1} &= g_{1,1}(x_1, x_2, x_3) & y_{2,1} &= g_{2,1}(x_1, x_2, x_3) \\
y_{1,2} &= g_{1,2}(x_1, x_2, x_3) & y_{2,2} &= g_{2,2}(x_1, x_2, x_3) \\
y_{1,3} &= g_{1,3}(x_1, x_2, x_3) & y_{2,3} &= g_{2,3}(x_1, x_2, x_3) \\
y_{1,4} &= g_{1,4}(x_1, x_2, x_3) & y_{2,4} &= g_{2,4}(x_1, x_2, x_3)
\end{aligned}
\tag{2.9}
$$

As illustrated in Figure 2.9, each of these is mapping the unit cube $E^3$ onto the unit interval, so that for every real-valued continuous function $f$ of four variables there are continuous functions $h_1, h_2, h_3, h_4$ of three variables for which

$$
f(x_1, \ldots, x_4) = \sum_{q=1}^{4} h_q(y_{1,q}, y_{2,q}, x_4)
\tag{2.10}
$$

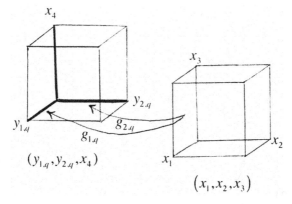

Figure 2.9: The Geometry of the Functions (2.9).

The cube $(x_1, x_2, x_3)$ is mapped onto intervals $y_{1,q}$ and $y_{2,q}$ respectively. Together with $x_4$ they constitute cubes $(y_{1,q}, y_{2,q}, x_4)$.

It is interesting to compare these three representations of functions of four variables. Keep in mind that the functions $g_q$ have values in the universal tree $\Xi$, whereas the functions $g_{pq}(x_1, x_2, x_3)$ are real valued.

1.  $f(x_1, ..., x_4) = \sum_{q=1}^{5} h_q(\xi_q) \quad \xi_q = g_q(x_1, x_2, x_3, x_4)$

2.  $f(x_1, ..., x_4) = \sum_{q=1}^{4} h_q(\xi_q, x_4) \quad \xi_q = g_q(x_1, x_2, x_3)$

3.  $f(x_1, ..., x_4) = \sum_{q=1}^{4} h_q(y_{1,q}, y_{2,q}, x_4) \quad y_{pq} = g_{pq}(x_1, x_2, x_3)$

It is also illuminating to compare their commuting diagrams, as illustrated in Figure 2.10.

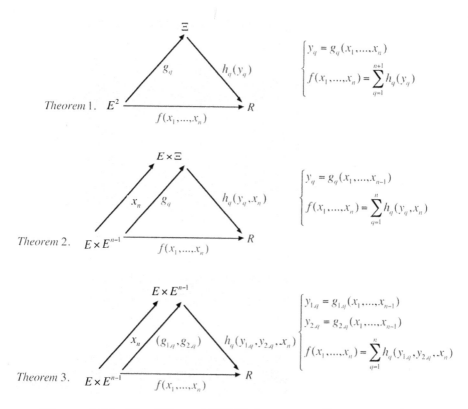

Figure 2.10: The Three 1956 Kolmogorov Representations

Figure 2.11 presents a flowchart of the representations (2.9)–(2.10). The four pairs of intervals form together with the interval $0 \leqslant x_4 \leqslant 1$ four new cubes $(y_{1,q}, y_{2,q}, x_4)$ for $q = 1,2,3,4$ on which the functions $h_q$ are defined and computed as follows:

A point $(x_1, x_2, x_3, x_4) = ((x_1, x_2, x_3), x_4)$ at which the functions $h_q$ are computed determines points $(y_{1,q}, y_{2,q}, x_4)$ through (2.9). The achieved outcome is that the function $f(x_1, \ldots, x_4)$ defined on a four-dimensional cube is computed with four functions $h_q$ each defined and computed on a three-dimensional cube.

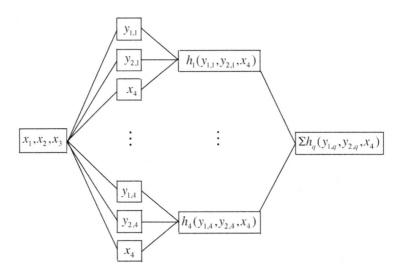

Figure 2.11: Flowchart of $f(x_1, ..., x_4) = \sum\limits_{q=1}^{4} h_q(y_{1,q}, y_{2,q}, x_4)$

Note that in the list of functions (2.9) the fixed functions $g_{1,q}$ and $g_{2,q}$ depend on $n$, so that their number of variables increases with $n$. But not so the functions $h_q$: these remain to be functions of three variables for any integer. This being the case, we wonder what would happen if we apply representations (2.9)–(2.10) to the functions

$$y_{1,q} = g_{1,q}(x_1, x_2, x_3, x_4)$$

$$y_{2,q} = g_{2,q}(x_1, x_2, x_3, x_4)$$

for the four values of $q$, and substitute the results into the upgraded equation (2.10) to:

$$f(x_1, ..., x_5) = \sum_{q=1}^{5} h_q(y_{1,q}, y_{2,q}, x_5)?$$

Reflecting on how we got here, we can actually guess what to expect: continuity will clearly be preserved, the functions

$$g_{1,q}(x_1, x_2, x_3, x_4)$$

and

$$g_{2,q}(x_1, x_2, x_3, x_4)$$

will be replaced by similar functions with only three variables, and the functions $h_q$ will remain to be functions of only three variables.

To carry this out, we get from (2.10) the representations:

$$g_{1,q}(x_1, x_2, x_3, x_4) = \sum_{q=1}^{4} \hat{h}_{1,q,r}(\hat{y}_{1,q,r}, \hat{y}_{2,q,r}, x_4)$$

$$g_{2,q}(x_1, x_2, x_3, x_4) = \sum_{q=1}^{4} \hat{h}_{2,q,r}(\hat{y}_{1,q,r}, \hat{y}_{2,q,r}, x_4)$$

where

$$\hat{y}_{1,q,r} = \hat{g}_{1,q,r}(x_1, x_2, x_3)$$

$$\hat{y}_{2,q,r} = \hat{g}_{2,q,r}(x_1, x_2, x_3).$$

Consequently (2.6) becomes

$$f(x_1, ..., x_5) =$$

$$\sum_{q=1}^{5} \left[ \sum_{r=1}^{4} \hat{h}_{1,q,r}(\hat{y}_{1,q,r}, \hat{y}_{2,q,r}, x_4), \sum_{r=1}^{4} \hat{h}_{2,q,r}(\hat{y}_{1,q,r}, \hat{y}_{2,q,r}, x_4), x_5 \right].$$

If we set further

$$z_{1,q} = \sum_{r=1}^{4} \hat{h}_{1,q,r}(\hat{y}_{1,q,r}, \hat{y}_{2,q,r}, x_4)$$

$$z_{2,q} = \sum_{r=1}^{4} \hat{h}_{2,q,r}(\hat{y}_{1,q,r}, \hat{y}_{2,q,r}, x_4)$$

then

$$f(x_1, ..., x_5) = \sum_{q=1}^{5} h_q(z_{1,q}, z_{2,q}, x_5) \qquad (2.11)$$

A side-by-side comparison of the terms

$$h_q(z_{1,q}, z_{2,q}, x_5)$$

and

$$h_q[g_{1,q}(x_1, x_2, x_3), g_{2,q}(x_1, x_2, x_3), x_4]$$

underlines the point that all functions involved are functions of three variables.

There is no conceptual difference between (2.10) and (2.11). It is simply that the sum increases and insertions become more complicated, but the crucial fact is that every stage involves only functions of three variables. This breakdown reveals the similarity between their geometries: a succession of mappings of three-dimensional cubes onto intervals that form the sides of other three-dimensional cubes.

The above demonstrates explicitly that continuous functions of four and five variables can be represented with superpositions of continuous functions of three variables, but now we can see how this process can be extended to continuous functions of any finite number of variables. Kolmogorov's *"rather unexpected consequence"* was that *an arbitrary real-valued continuous function of four or more variables can be represented with superposition of real-valued continuous functions of three variables!*

The actual statement of this breakthrough is this:

**Theorem 3.** *For any integer $n \geqslant 3$ there exist real-valued continuous functions*

$$y_{1,1} = g_{1,1}(x_1, x_2, x_3) \quad y_{2,1} = g_{2,1}(x_1, x_2, x_3)$$

$$\vdots \qquad\qquad\qquad \vdots$$

$$y_{1,n} = g_{1,n}(x_1, x_2, x_3) \quad y_{2,n} = g_{2,n}(x_1, x_2, x_3)$$

*defined on $E^n$, for which every real-valued continuous function $f$ defined on $E^n$ has a representation*

$$f(x_1, ..., x_n) = \sum_{q=1}^{n} h_q(y_{1,q}, y_{2,q}, x_n)$$

*with continuous functions $h_q$.*

It is important to remember that, like the other results that had been discussed so far, this is another existence theorem. It was not intended as a computational tool, and no actual examples of such representations exist. This situation will keep haunting us until flesh is added in later chapters to the bones of these constructions.

We understand now why Kolmogorov's strategy did not allow a better lower bound on the number of variables in the superpositions. This result could not refute Hilbert's specific conjecture for functions of three variables in terms of representations with functions of two variables, but even with its limitations this was an extraordinary result.

Setting this limitation aside, Kolmogorov breached the dike, so to speak, by establishing that there were no continuous functions of more than three variables, and he came within a hairbreadth from disproving Hilbert's conjecture. It is clear already at this stage that this conjecture no longer holds central stage; that superpositions and function representations developed far beyond Hilbert's problem.

That solving Hilbert's Problem 13 was indeed the goal of this research, was stated by Kolmogorov in the second paragraph of this paper:

> The question of the possibility of representing an arbitrary continuous of two variables remains open. The proof of the possibility of such a representation would yield a complete solution of Hilbert's $13^{th}$ problem in the sense of establishing the hypothesis stated by Hilbert.

The complete statement is quoted because of its importance in stating unambiguously the goal, and qualifying it. Namely, that this attempt was aimed at solving the narrow conjecture and not the part dealing with superpositions of functions having smoothness conditions. This reading of the statement is supported by the communication (Kolmogorov [56]) and elsewhere.

Importantly, it followed from Kolmogorov's finding that the worst continuous functions of any finite number of variables were no worse than continuous functions of three variables. As mentioned earlier, this ran counterintuitive to the notion that complexity should increase with the degree of freedom that higher dimensions provide in some quantifiable sense.

The strategy that led to this breakthrough was highly innovative and moved conceptually and technically away from any hint of algebra and the arsenal of tools employed at the first third of the twentieth century.

The 1956 paper contained yet another overlooked result: that every real-valued continuous function defined on the unit cube can be approximated with polynomials of two variables. In Kolmogorov's words, this "illuminates from a rather new side the circle of problems relating to Hilbert's $13^{th}$ problem."

This is the point at which Kolmogorov handed the gauntlet to Arnold, so to speak. Arnold did not use "the unexpected result" as his starting point, and instead backtracked and focused on the case $n = 3$ of formula in Theorem 2 from which it was derived.

We already know that this formula did not apply to Hilbert's conjecture because the continuous functions had values in the universal tree $\Xi$: $g_q : E^2 \to \Xi$, and the functions had values in the cross-product of $\Xi$ and the unit interval $h_q : \Xi \times E \to R$. These were therefore not real-valued continuous functions as stipulated in the conjecture.

Arnold observed that the universal tree could be replaced by a tree in $E^3$ with a branching index not exceeding 3 with appropriate modifications in Kolmogorov's construction. With this, he obtained the representation formula that was featured in the preface. Arnold's finding was stated in

**Theorem 4.** *There exist nine real-valued continuous functions*

$$y_1 = \varphi_1(x_1, x_2)$$

$$\vdots$$

$$y_2 = \varphi_2(x_1, x_2)$$

$$y_9 = \varphi_9(x_1, x_2)$$

(2.12)

*defined on the unit square $E^2$, for which every real-valued continuous function $f$ defined on $E^3$ has a representation*

$$f(x_1, x_2, x_3) = \sum_{q=1}^{9} g_q(y_q, x_3)$$

*with continuous functions $g_q$.*

This clearly refutes directly Hilbert's conjecture and more. Recall that Kolmogorov derived from his function representations a formula of this type with only seven summands.

We again chose this format instead of the compact version

$$f(x_1, x_2, x_3) = \sum_{q=1}^{9} g_q[\ \varphi_q(x_1, x_2), x_3]$$

to show that this representation depends on the over-determined system of equations (2.12). Having more equations than variables means that the two independent variables $x_1, x_2$ determine a relation between the nine variables $y_1, ..., y_9$ that are therefore not independent of each other. These variables can be regarded as variables in a nine-dimensional space, varying in a nine-dimensional cube $E^9$. Eliminating $x_1, x_2$ from this system would therefore determine a two-dimensional surface in this cube. We illustrate this with an example.

Let

$$y_1 = x_1 - x_2$$

$$y_2 = x_1 + x_2$$

$$y_3 = 4x_1 x_2 + 1$$

$$f(x_1, x_2) = g_1(y_1) + g_2(y_2) + g_3(y_3)$$

Like (2.12), this system of equations is over determined, and eliminating the variables $x_1$ and $x_2$ gives the equation:

$$y_3 = y_2^2 - y_1^2 + 1.$$

The graph of this equation is the saddle shown in Figure 2.12; it is a 2-dimensional surface in 3-dimensional space. To every point $(x_1, x_2) \in E^2$ corresponds a point $(y_1, y_2, y_3) \in D^3$, where for example, the point $(x_1, x_2) = (2, 1)$ corresponds to the point

$(y_1, y_2, y_3) = (1, 3, 9)$ on this surface. The line $x_1 = x_2$, for example, determines the parabola $y_3 = y_2^2 + 1$ along which the superposition would be calculated. In this example we would ask which functions can be represented with superpositions:

$$f(x_1, x_2) = g_1(x_1 - x_2) + g_2(x_1 + x_2) + g_3(4x_1x_2 + 1).$$

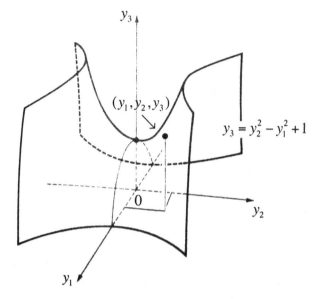

Figure 2.12: The Saddle $y_3 = y_2^2 - y_1^2 + 1$.

Arnold's study of finite trees in $E^3$ with branching index not exceeding three was an insightful and crucial change in Kolmogorov's construction. Combined with the 1956 paper, it showed him how to eliminate $\Xi$ while also reducing the lower bound of the number of variables from 3 to 2. It is for this reason that some researchers refer to Kolmogorov's formula as the Arnold - Kolmogorov formula. Perhaps under different circumstances, Kolmogorov's 1957 paper might have been a collaborative publication.

To understand Arnold's representation, we examine the system of equations composed of the inner functions (2.12) that are independent of $f$ Each of the nine functions is defined on the unit square that it maps onto the interval $E$. Each function $g_q$ is defined on the Cartesian product (instead of as was the case earlier), and repeating here the discussion that we used in Kolmogorov's third representation: Since $E^3 = E^2 \times E$, we can think of $f$ as being defined on $E^2 \times E$, and writing $f(x_1, x_2, x_3) = f((x_1, x_2), x_3)$ to describe it schematically with the commuting diagram in Figure 2.13.

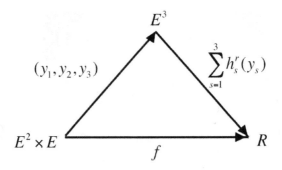

Figure 2.13: Commuting Diagram

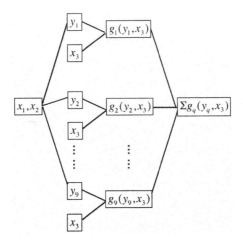

Figure 2.14: Flowchart of $f(x_1, x_2, x_3) = \sum_{q=1}^{9} g_q[y_q, x_3]$

The intermediate mapping from $E^2 \times E$ to $E^3$ is independent of $f$, and as the flowchart shows in Figure 2.14 shows, the computation of $f$ is a two-stage process, first replacing the ordered pair s $(x_1, x_2)$ with the variables $y_1, \ldots, y_9$, and then computing $f$.

Talking forty years later about this work in mathematics, Arnold made light of this achievement:

> Kolmogorov had proved that it is sufficient to represent any function on a tree in Euclidean space—actually, to find a universal tree such that any continuous function on this tree can be represented as the sum of three continuous functions, each depending on one coordinate. If you can do this, then "there are no functions of three variables" and you can reduce any continuous to continuous functions of two variables—and the function $z(a, b, c)$ is reducible too. This was a problem I managed to solve. It was essentially simple. ... In the simplest example of Figure 2.2 the claim is that on this tree any function can be represented as $f(x) + g(y)$. How can you do this? (Arnold [5])

Figure 2.2 that Arnold referred to is a stripped version of our Figure 2.5 above. He presents an argument similar to the one accompanying that figure:

> That's all. If the tree is more complicated you will have more branches or even an infinite number of branches, you will have to work more, and in fact to make the infinite process converge you will need three variables not two, but that's the principal idea—that's how it works. (*Ibid.*)

How much more would we have to work? Arnold's detailed proof of his Doklady paper is 72 pages in length, including an appendix with some essential background information (Arnold [4]).

This is a very interesting paper in its own right, accessible to the reader without prior knowledge of the subject; for the uninitiated, it would be worthwhile to read the appendix first.

We are now on the threshold of Kolmogorov's seminal and ultimate discovery in the area of classification of continuous functions. A more detailed discussion of trees is beyond the scope of our considerations, and they will play no further role in this manuscript. But the details of this approach had not been the point for including it of our story: It was the strategy of linearizing the domain, so to speak, rather than dealing with functions directly. In this Kolmogorov's Function Representations Theorem was a complete break with earlier strategies, and this was the road to it.

## 2.4  Kolmogorov's Function Representations

Our story brings us now to the centerpiece of this monograph: Kolmogorov's Representation Theorem (Kolmogorov [55]). A summary of a few earlier observations and comments is presented here to avoid an otherwise necessity to leaf through earlier pages of the narrative. This begins with the by now familiar format for representation of functions of two variables:

$$y_1 = \psi_{1,1}(x_1) + \psi_{2,1}(x_2)$$

$$y_2 = \psi_{1,2}(x_1) + \psi_{2,2}(x_2)$$

$$y_3 = \psi_{1,3}(x_1) + \psi_{2,3}(x_2) \qquad (2.13)$$

$$y_4 = \psi_{1,4}(x_1) + \psi_{2,4}(x_2)$$

$$y_5 = \psi_{1,5}(x_1) + \psi_{2,5}(x_2)$$

$$f(x_1, x_2) = \sum_{q=1}^{5} g_q(y_q)$$

It highlights the prime importance of the functions $\psi_{pq}(x)$ in the representations, the uniformity of its construction and the additive feature of the superpositions; it also makes it easier to see how it would be implemented: A point $(x_1, x_2) \in E^2$ generates five numbers $y_1, y_2, y_3, y_4, y_5$ at which the functions $g_1, g_2, g_3, g_4, g_5$ compute a target function $f$.[4]

Observe that this execution alternates between nonlinear and linear computations:[5]

$$\left[\psi_{1,q}(x_1), \psi_{2,q}(x_2)\right] \rightarrow \left[\psi_{1,q}(x_1) + \psi_{2,q}(x_2)\right]$$

computing functions, $q$=1,2,3,4,5     adding functions, $q$=1,2,3,4,5

nonlinear         linear

$$\rightarrow \left[g_1(y_1), ..., g_5(y_5)\right] \rightarrow \left[g_1(y_1) + ... + g_5(y_5)\right]$$

computing functions     adding functions

nonlinear        linear

Figure 2.15: The Architecture of Kolmogorov's Formula

[4]See Figure 3 in the Preface.
[5]See Figure 2.15.

An over-determined systems of transfer equations inherent in dimension-reduction formulas is the case also with the system (2.13) the variables $y_1, ..., y_5$ are related to each other through the variables $x_1, x_2$. They can be regarded as variables in a five-dimensional space, and eliminating $x_1, x_2$ from the system (2.13) would give a two-dimensional surface in five-dimensional space. This was illustrated in the above example.

Recall our first discussion of commuting diagrams. The diagram in Figure 2.16 explains the superposition mappings.

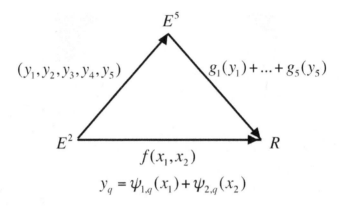

Figure 2.16: Commuting Diagram of Kolmogorov's Superpositions

It seems appropriate to present the actual formulation of Kolmogorov's theorem. With the exception of slightly different notation to conform to usage in this monograph, here is the verbatim text of the translation of H. P. Thielman of Kolmogorov's theorem It appeared in the 1963 issue of the American Mathematical Society Translations

**Kolmogorov's function representation's Theorem.** *For every $n \geqslant 2$ there exist continuous real functions $\psi_{pq}(x)$, defined on the unit interval $E = [0, 1]$, such that every continuous real function $f(x_1, ..., x_n)$, defined on the $n$-dimensional cube $E^n$, is representable*

*in the form*

$$f(x_1, ..., x_n) = \sum_{q=1}^{2n+1} g_q \circ \sum_{p=1}^{n} \psi_{pq}(x_p)$$

*where the functions $g_q(y)$ are real and continuous.*

In Kolmogorov's words:

> [T]he constructions in this note were found by analyzing the constructions used in [his 1956 paper and Arnold's paper] and by eliminating the details that were superfluous for obtaining the final result.

This explains both the brevity of this groundbreaking paper, and also the view of some researchers that this was in the nature of a collaborative effort.

Continuing with the two dimensional case, we already observed that the connection between the function $f$ being represented and the functions $g_q$ that compute it is through an intermediate step: the transformation of the square $E^2$ into a 2-dimensional surface in 5- dimensional space where the computation is carried out. This suggests that the characteristics of this surface are indeed critical.

Kolmogorov's conceived of the systems of functions (2.13) geometrically, as one can also glean from his approach in the 1956 paper. He outlined the strategy that we used in constructing the relevant functions. Assuming the inner functions available, he proved inductively how to approximate the functions and simultaneously the target function $f$ in question. This inductive proof depends on a certain property of the functions $\psi_{1,q}(x_1) + \psi_{2,q}(x_2)$ for which specific format is not relevant. It will therefore be convenient to rephrase the conditions in Kolmogorov's theorem in terms of continuous functions $\varphi_q(x_1, x_2)$. This was the opening statement of this monograph.

Proposition (1) with which the narrative opened summarized two remarkably simple requirements that these dimension-reducing representations: $2n + 1$ families of pairwise disjoint closed $n$-dimensional cubes with diameters that diminish to zero cover the unit cube $E^n$ in a certain way; and $2n + 1$ continuous functions map each family of cubes into pairwise disjoint intervals. We restate this proposition in a formal language.

## 2.5 Function Representations

**Theorem 5.** *For a given integer $n \geqslant 2$ let*

$$\{S_{1,k}\}, \{S_{2,k}\}, ..., \{S_{2n+1,k}\}, k = 1, 2, 3, ...$$

*be $2n+1$ five families of closed $n$-dimensional cubes with diameters diminishing to zero as $k$ tends to infinity, such that:*

a) *Every point belongs to at least $2n$ cubes $S_{q,k}$ for each $k$;*

b) *$S_{q,k} \cap S_{q,k'} = \emptyset$ whenever $S_{q,k} \neq S_{q,k'}$ for each $q$ and $k = 1, 2, 3 ...$*
   *If the $2n + 1$ continuous functions $y_q = \varphi_q(x_1, ..., x_n)$, $q = 1, 2, ..., 2n + 1$ that separate the cubes $S_{q,k}$ for each $q$ and $k = 1, 2, 3 ... :$*

c) *$\varphi_q(S_{q,k}) \cap \varphi_q(S_{q,k'}) = \emptyset$ whenever (b) holds,*

*then every real-valued continuous function $f(x_1, ..., x_n)$ defined on $E^n$ has a representation*

$$f(x_1, ..., x_n) = \sum_{q=1}^{2n+1} g_q \circ \varphi_q(x_1, ..., x_n)$$

*with continuous functions $g_q$.*[6]

---

[6]See Fridman [35] and Sprecher [9].

Note that the crucial and only requirement for each of the functions $\varphi_q$ is that all cubed have non-intersecting images for each value of $k$. The images of cubes are intervals on the real line, and this separation of intervals leads to a separation of points property.[7] For simplicity we shall limit constructions to the case $n = 2$.

In this particular scheme, one family of squares ($q = 1$) is translated four times along the main diagonal. All squares are of equal dimensions for each value of $k$, and the separations between them are uniform. In the depiction of Figures 1 and 2, the lower left corners are aligned with the upper right corners with increasing $q$, so that each successive translations cover as much as possible of the gaps between an underlying family.

In this geometry, small corner squares are left exposed after each translation. This is why two translations (three families) are required to provide a single complete covering. We shall actually construct below families of squares $S_{q,k}$ meeting these conditions.

We note in passing that this scheme is based on the covering theorem of Lebesgue that tells us that three families (two translations) are required to cover every point of a square at least once. The reader can deduce from the figure that with three families some points would actually be covered exactly once. A total of five families (four translations) is necessary to guarantee an eventual convergence of the approximating functions $g_q$ to the target function $f$ as we shall show below.

Conditions (b) and (c) in Theorem 5 imply that squares and intervals respectively corresponding to the same value of $k$ are either coincident (contain exactly the same points) or else have no points in common. This must remain so as the diameters of the squares and intervals diminish to zero.

Among the hidden but essential requirements of all function representations is the point separation property. Reverting to the case n=2, this means the following: For any two points $(x_1, x_2) \neq$

---

[7]See Figure 1 and 2

$(x_1', x_2')$ of $E^2$ there is a continuous function $f$ that has different values at these particular points: $f(x_1, x_2) \neq f(x_1', x_2')$.

The reader should be able to come up with any number of examples for which this is true. Thus:

$$\varphi_1(x_1, x_2) + \ldots + \varphi_5(x_1, x_2) \neq \varphi_1(x_1', x_2') + \ldots + \varphi_5(x_1', x_2')$$

and equivalently

$$[\varphi_1(x_1, x_2) - \varphi_1(x_1', x_2')] + \ldots + [\varphi_5(x_1, x_2) - \varphi_5(x_1', x_2')] \neq 0$$

This implies that

$$[\varphi_q(x_1, x_2) - \varphi_q(x_1', x_2')] \neq 0$$

for at least one value of $q$. This implies, in turn, that all points of $E^2$ must somehow be separated by the functions $\varphi_q(x_1, x_2)$ in this sense. This tells us actually more. Recall that the over determined system of equations specifies a two-dimensional surface in five-dimensional space, implying that the images of any two points $(x_1, x_2) \neq (x_1', x_2')$ of $E^2$ on this surface do not coincide:

$$(y_1, \ldots, y_5) \neq (y_1', \ldots, y_5')$$

This is illustrated in Figure 2.17.

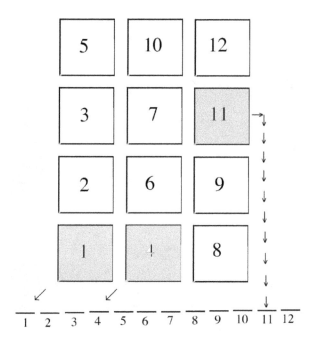

Figure 2.17: A function $\varphi_q(x_1, x_2)$ for some $q$. Pairwise disjoint squares are mapped onto pairwise disjoint pairwise disjoint intervals in an order determined by $\varphi_q(x_1, x_2)$.

We now construct families of squares $S_{q,k}$ as translates of a single family as depicted in Figure 1 meeting the conditions above. These in turn are the Cartesian-products of intervals laid along the coordinate axes of $E^2$. Because of the required covering requirement, the width of the gaps cannot exceed one-fifth the length of intervals. Using base 10, we use decimals

$$d_k = \sum_{r=1}^{k} \frac{i_r}{10^r},$$

where $i_1 = 0, 1, ..., 10$ and $i_r = 0, 1, ..., 9$ for values $r > 1$ and define for each value of $k$ intervals of length

$$\delta_k = \frac{8}{9 \cdot 10^k} = 8 \sum_{r=k+1}^{\infty} \frac{1}{10^r}$$

separated by gaps of width $\frac{1}{9}10^{-k}$; these intervals are

$$E_k(d_k) = [d_k, d_k + \delta_k] \qquad (2.14)$$

and they are illustrated in Figure 2.18 together with the relation between $E_k(d_k)$ and $E_{k+1}(d_{k+1})$.

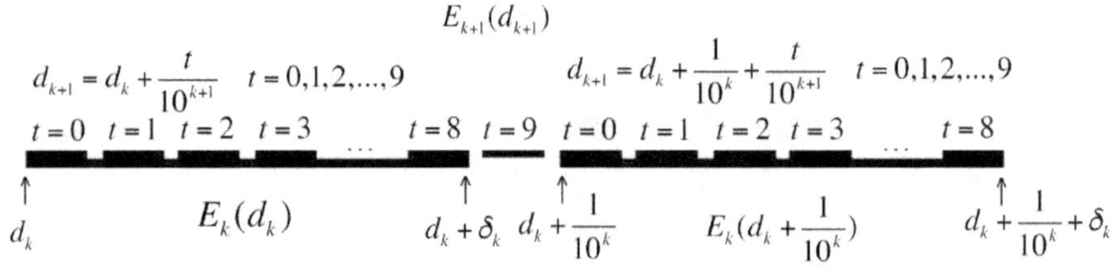

Figure 2.18: Intervals $E_k(d_k)$ and $E_{k+1}(d_{k+1})$

In taking an inventory of the properties of these intervals, note that the initial points of intervals $E_{k+1}(d_{k+1})$ coincide with initial points of intervals $E_k(d_k)$ when $i_{k+1} = 10i_k$, and so do terminal points coincide when $i_{k+1} = 10i_k + 8$ as shown in the figure. When $i_{k+1} = 10i_k + 9$, the interval lies in the gaps between the intervals $E_k(d_k)$ and $E_k(d_k + \frac{1}{10})$. Consequently,

$$E_k(d_k) \cap E_{k+1}(d_{k+1}) \neq \emptyset, \qquad (2.15)$$

for $i_{k+1} = 10i_k + t$ and $t = 0, 1, ..., 8$, and $E_{k+1}(d_{k+1})$ is contained in a gaps when $t = 9$. Also,

$$\frac{i_k}{10^k} + \delta_k = \frac{10i_k + 8}{10^{k+1}} + \delta_{k+1}$$

for each value of $i_k$, and it follows that if (2.14) holds, then[8]

$$E_k(d_k) \supset E_{k+1}(d_{k+1}).$$

For each $k$, we translate these intervals by distances

$$\frac{q}{9 \cdot 10}, q = 0, 1, 2, 3, 4$$

and define the new families of intervals

$$E_{q,k}(d_k) = \left[ d_k - \frac{q}{9 \cdot 10}, d_k + \delta_k - \frac{q}{9 \cdot 10} \right]. \tag{2.16}$$

These numbers are selected so that the initial points of intervals $E_{q,k}(d_k)$ coincide with the endpoints of intervals $E_{q+1,k}(d_k)$ as illustrated in the figure.

$q=5$
$q=4$
$q=3$
$q=2$
$q=1$

Figure 2.19: Translated intervals $E_{q,k}(d_k)$ for fixed $k$

For fixed $q$, the intervals $E_{q,k}(d_k)$ are also separated by gaps of width $\frac{1}{9}10^{-k}$; each interval $E_{q,k}(d_k)$ contains eight intervals $E_{q,k+1}(d_{k+1})$ and each gap contains one such interval, and this tells us all we need to know about the geometry of these intervals.

Note: To cover the unit interval $E$ for each value of $q$, the rational numbers $d_k$ must be determined for each interval $E_{q,k}(d_k)$ to compensate for the translations $-\frac{q}{9 \cdot 10}$. We prefer not to do this in order not to add further complication to the notation. It will always be assumed that the last interval for each value of $q$ contains the end point $(1,0)$ of the unit interval.

---

[8]See Figure 2.18.

Before moving on, there is something else that we can learn from examining Figure 2.18. The first observation is that the picture does not change with changing $k$ other than its scale. The placement of nine $(k+1)$-intervals in each k-interval for $k = 1, 2, 3, \ldots$ results in nested intervals whose lengths tend to zero and they converge to a point. Since every gap contains such nested intervals as well, we deduce that the set of points generated this way is dense everywhere in $E$. This implies that any open interval in $E$ no matter how small contains infinitely many points generated through these nested interval. Perhaps not as obvious, but with a little imagination you will discover further that the values of $k$ need not be consecutive to get a nested sequence of intervals.

Next, we define the Cartesian products:

$$S_{q,k}(d_{1,k}, d_{2,k}) = E_{q,k}(d_{1,k}) \times E_{q,k}(d_{2,k}), \qquad (2.17)$$

for $q = 1, 2, 3, 4, 5$ and $k = 1, 2, 3, \ldots$

These squares have diameters $\sqrt{2} \cdot \delta_k$, and it follows at once that they tend to zero as $k \to \infty$; likewise the diameters $\frac{\sqrt{2}}{9} 10^{-k}$ of the gaps separating them tend to zero as $k \to \infty$.

Since the largest translation along any coordinate axis is $\frac{4}{9} 10^{-k}$, it is easily verified that that each point of $E^2$ is contained in the intersection of at least 3 squares $S_{q,k}(d_{1,k}, d_{2,k})$.

These families of squares satisfy properties (a), (b), and (c) of Theorem 5 and this completes the first step of the construction. We are ready to construct an appropriate continuous function $\psi(x)$ that can be used to satisfy property (d).

# Chapter 3

# Superpositions

## 3.1 The function $\psi(x)$

Function representations with sums of superpositions of the form $g_q[\varphi_q(x_1, ..., x_n)]$ are not difficult to come by. The heart of Kolmogorov's function representations are the functions $\psi_{pq}(x)$ whose sums:

$$y_q = \psi_{1,q}(x_1) + \psi_{2,q}(x_2) + ... + \psi_{n,q}(x_n)$$

have the remarkable point-separating property of Theorem 5. These functions set Kolmogorov's representations apart from everything that preceded them or came since. To this author, their discovery rather than that of the functions $g_q(y_q)$ represent Kolmogorov's stroke of brilliance.

He had already the substance of convergence schemes developed the establishment of dimension-reducing representations in his 1956 paper, but these have been indirect and cumbersome. Kolmogorov's great success was the discovery of real-valued functions $\psi_{pq}(x)$ that could eliminate the universal tree $\Xi$ as an intermediary agency.

The construction and anatomy of the functions $\psi_{pq}(x)$ is the exclusive focus of attention at this point, though actually the honors will go to a single surrogate function$\psi(x)$. This function and its translates $\psi(x + qa)$ were already encountered in the preface in Robert Hecht-Nielsen's neural network. This chapter begins, therefore, with a heuristic argument to explain why the functions $\psi_{pq}(x)$ are equivalent to the translates $\psi(x+qa)$of a single function We argue this for the case $n = 2$.

Toward this end, let $q$ be fixed and set

$$\varphi_q(x_1, x_2) = \psi_{1,q}(x_1) + \psi_{2,q}(x_2)$$

condition **B** of Theorem 5 implies that

$$\psi_{1,q}(d_{1,k}) + \psi_{2,q}(d_{2,k}) \neq \psi_{1,q}(d'_{1,k}) + \psi_{2,q}(d'_{2,k}) \qquad (3.1)$$

when $(d_{1,k}, d_{2,k}) \neq (d'_{1,k}, d'_{2,k})$ for all values of $k$. This is an essential condition of Kolmogorov's constructions.

If we define a function $\psi(x)$ whose values $\psi(d_k)$ are rational for all rational numbers and for all values of $k$, then

$$\psi(d_{1,k}) + \lambda\psi(d_{2,k}) \neq \psi(d'_{1,k}) + \lambda\psi(d'_{2,k})$$

for any irrational number $\lambda$ whenever $(d_{1,k}, d_{2,k}) \neq (d'_{1,k}, d'_{2,k})$.

Furthermore, this remains true for translates:

$$\psi(d_{1,k} + qa) + \lambda\psi(d_{2,k} + qa) \neq \psi(d'_{1,k} + qa) + \lambda\psi(d'_{2,k} + qa) \quad (3.2)$$

when $(d_{1,k}, d_{2,k}) \neq (d'_{1,k}, d'_{2,k})$ for any rational constant $a > 0$.

Instead of (3.2) we shall therefore consider the representation

$$y_0 = \psi(x_1) + \lambda\psi(x_2)$$

$$y_1 = \psi(x_1 + a) + \lambda\psi(x_2 + a)$$

$$y_2 = \psi(x_1 + 2a) + \lambda\psi(x_2 + 2a) \qquad (3.3)$$

$$y_3 = \psi(x_1 + 3a) + \lambda\psi(x_2 + 3a)$$

$$y_4 = \psi(x_1 + 4a) + \lambda\psi(x_2 + 4a)$$

$$f(x_1, x_2) = \sum_{q=0}^{4} g_q(y_q)$$

for suitable constants $\lambda$ and $a$, subject to showing that the images $\psi(S_{q,k}(d_{1,k}, d_{2,k}))$ do not intersect for each $q$ and all grid-points.

Like the functions $\psi_{pq}(x)$, the function $\psi(x)$ will be obtained as the uniform limit of devil's-staircase type functions $\psi_k(x)$ that Kolmogorov devised.[1] We construct them geometrically using the numbers

$$\beta(k) = 2^k - 1 = 1 + 2 + 2^2 + ... + 2^{k-1} \tag{3.4}$$

and intervals defined in (2.14).

We follow this lead-in with simple geometric graphs and some involved arithmetic, beginning with the definition of the point-function

$$\psi(d_1) = d_1 \text{ for } i_1 = 0, 1, 2, ..., 10$$

Its graph consists of equally spaced point $(d_1, d_1)$ on the main diagonal.

The point-function $\psi_2$ that is defined next keeps these graph points, and adds points between them. It is first defined on the interval $\left[0, \frac{1}{10}\right]$:

$$\psi_2\left(\frac{i_2}{10^2}\right) = \begin{cases} \frac{i_2}{10^{\beta(2)}}, & i_2 = 0, 1, 2, ..., 8 \\ \frac{1}{2}\left[\frac{8}{10^{\beta(2)}} + \frac{1}{10}\right], & i_2 = 9 \end{cases} \tag{3.5}$$

$$\psi_2\left(\frac{1}{10}\right) = \frac{1}{10}$$

as depicted in Figure 3.1. Note that the graph no longer consists of evenly spaced points.

---

[1] See Figures 3.1, 3.2, and 3.3 below.

58

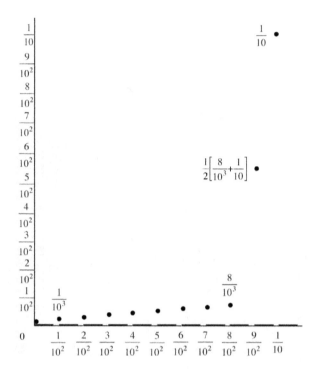

Figure 3.1: The function $\psi_2\left(\frac{i_2}{10^2}\right)$ on the interval $\left[0, \frac{1}{10}\right]$.

A staircase function $\psi_2(x)$ is constructed on this interval by connecting these points with line segments as shown in Figure 3.2. That is:

$$\psi_2(x) = \frac{i_2}{10^{\beta(2)}} \text{ when } x \in E_2\left(\frac{i_2}{10^2}\right) \text{ and } i_2 = 0, 1, 2, ..., 8,$$

$$\psi_2(x) = \frac{1}{2}\left[\frac{8}{10^{\beta(2)}} + \frac{1}{10}\right] \text{ when } x \in E_2\left(\frac{9}{10^2}\right),$$

and the resulting steps are connected with sloping line segments.

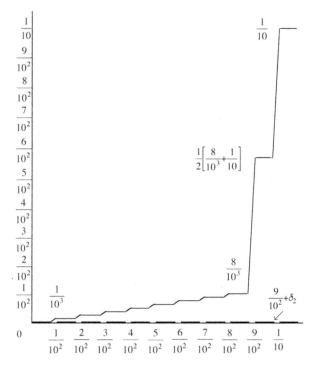

Figure 3.2: The function $\psi_2(x)$ on the interval $\left[0, \frac{1}{10}\right]$.

The function $\psi_2(x)$ is extended to the interval $\left[0, 1 + \frac{1}{10}\right]$ by setting

$$\psi_2\left(\frac{i_1}{10} + \frac{i_2}{10^2}\right) = \frac{i_1}{10} + \psi_2\left(\frac{i_2}{10^2}\right).$$

This results in the formula:

$$\psi_2\left(\frac{i_1}{10} + \frac{i_2}{10^2}\right) = \psi_1\left(\frac{i_1}{10}\right) + \begin{cases} \frac{i_2}{10^{\beta(2)}}, & i_2 = 0, 1, 2, ..., 8 \\ \frac{1}{2}\left[\frac{8}{10^{\beta(2)}} + \frac{1}{10}\right], & i_2 = 9 \end{cases} \tag{3.6}$$

Likewise, set

$$\psi_2\left(x + \frac{i_1}{10}\right) = \psi_2\left(x\right) + \frac{i_1}{10}$$

for values $i_1 = 0, 1, 2, ..., 10, 0 \leqslant x \leqslant \frac{1}{10}$.

This gives the graph in Figure 3.3. I hope that no confusion will arise from using the symbol $\psi_2$ to designate both the pointwise function and continuous staircase functions.

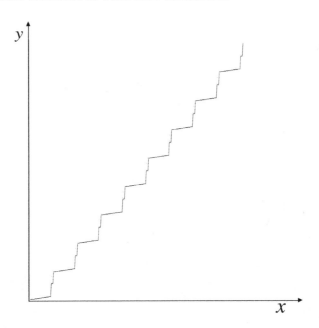

Figure 3.3: Partial graph of the second approximation $\psi_2(x)$

Kůrková made the interesting suggestion that Kolmogorov's use of devil staircase-type functions may have been inspired by 19th century staircase functions. A heuristic argument to explain the transition from trees to devil's staircase functions might go like this: Think of a finite tree and a path up a piecewise linear branch. This zigzag may have branches with positive slopes, and others with negative slopes. One can imagine replacing negative slopes with their mirror images, and transforming any such zigzag into a strictly monotonic piecewise linear path with no negative slopes.

The graphs in Figure 3.2 and 3.3 can be regarded as prototypes.

Looking ahead, the construction of point-functions

$$\psi_3(d_3), \psi_4(d_4), \psi_5(d_5), ...$$

and the attendant continuous devil staircase functions

$$\psi_3(x), \psi_4(x), \psi_5(x), ...$$

can be realized by superimposing appropriately scaled copies of Figure 3.1 and Figure 3.2 into the decreasing intervals

$$\left[\frac{i_k}{10^k}, \frac{i_k + 1}{10^k}\right]$$

Each of the functions $\psi_k(x)$ so constructed will be continuous and monotonic increasing, and so will be limit function $\psi(x) = \lim_{k \to \infty} \psi_k(x)$.

To show the recursive nature of the construction, let us push the proverbial envelope another notch by developing an explicit formula for the point-function

$$\psi_3(d_3) = \psi_3\left(\frac{i_1}{10} + \frac{i_2}{10^2} + \frac{i_3}{10^3}\right)$$

We define the function $\psi_3(d_3)$ in stages: first

$$\psi_3\left(\frac{i_3}{10^3}\right) \text{ on the interval } \left[0 + \frac{1}{10^2}\right],$$

$$\psi_3\left(\frac{i_2}{10^2} + \frac{i_3}{10^3}\right) \text{ on the interval } \left[0 + \frac{1}{10}\right]$$

and finally $\psi_3(d_3)$.

Here is the first step:[2]

$$\psi_3\left(\frac{i_3}{10^3}\right) = \begin{cases} \frac{i_3}{10^{\beta(3)}}, & i_3 = 0,1,2,...,8 \\ \frac{1}{2}\left[\frac{8}{10^{\beta(3)}} + \frac{1}{10^{\beta(2)}}\right], & i_3 = 9 \end{cases} \qquad (3.7)$$

$$\psi_3\left(\frac{1}{10^2}\right) = \frac{1}{10^{\beta(2)}}$$

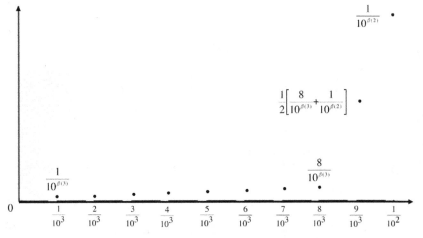

Figure 3.4: The function $\psi_3\left(\frac{i_3}{10^3}\right)$ on the interval $\left[0, \frac{1}{10^2}\right]$.

Note the similarity in geometry between $\psi_2\left(\frac{i_2}{10^2}\right)$ and $\psi_3\left(\frac{i_3}{10^3}\right)$. Formula (3.6) guides the construction of $\psi_3\left(\frac{i_2}{10^2} + \frac{i_3}{10^3}\right)$ on the interval $\left[0, \frac{1}{10}\right]$. There is no difference in the pattern between this and $\psi_2\left(\frac{i_1}{10} + \frac{i_2}{10^2}\right)$; computations remain straight forward, but the

---

[2]See Figure 3.4.

branching becomes increasingly convoluted.

$$\psi_3\left(\frac{i_2}{10^2}+\frac{i_3}{10^3}\right)=\begin{cases}\frac{i_2}{10^{\beta(2)}}+\begin{cases}\frac{i_3}{10^{\beta(3)}},\ i_3=0,1,2,...,8\\[4pt]\frac{1}{2}\left[\frac{8}{10^{\beta(3)}}+\frac{1}{10^{\beta(2)}}\right],\ i_3=8\end{cases}\quad i_2=0,1,...,8\\[14pt]\frac{1}{2}\left[\frac{8}{10^{\beta(2)}}+\frac{8}{10^{\beta(3)}}+\frac{1}{2}\left(\frac{1}{10^{\beta(2)}}+\frac{1}{10}\right)\right],\ i_2=8,\ i_3=9\\[10pt]\frac{1}{2}\left[\frac{8}{10^{\beta(3)}}+\frac{1}{10^{\beta(2)}}\right],\ i_2=9\ i_3=0\\[10pt]\frac{1}{2}\left\{\frac{1}{2}\left[\frac{8}{10^{\beta(3)}}+\left(\frac{1}{10^{\beta(2)}}+\frac{1}{10}\right)\right]+\frac{1}{10}\right\},\ i_2=9\ i_3=9\end{cases}$$

$$\psi_3\left(\frac{1}{10}\right)=\frac{1}{10}.$$

A detail of this function is shown in Figure 3.5. The function

$$\psi_3\left(\frac{i_1}{10}+\frac{i_2}{10^2}+\frac{i_3}{10^3}\right)$$

follows by setting

$$\psi_3\left(\frac{i_1}{10}+\frac{i_2}{10^2}+\frac{i_3}{10^3}\right)=\psi_1\left(\frac{i_1}{10}\right)+\psi_3\left(\frac{i_2}{10^2}+\frac{i_3}{10^3}\right)$$

and from this $\psi_3(x)$ is obtained the way we obtained $\psi_1(x)$.

As can be expected, each of the 'jumps' in this figure is going to be decreased by a little more than one half in the next iteration, and so on.

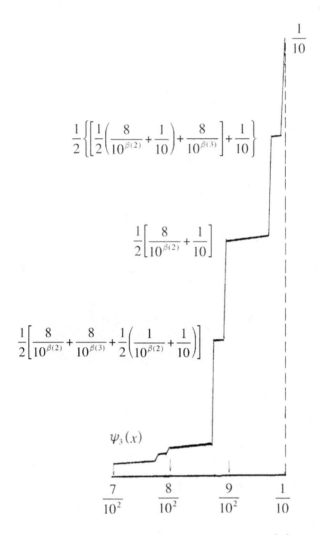

Figure 3.5: Detail of the function $\psi_3(x)$.

Using the decimal base in constructing the functions was a mere convenience because the calculations are easily carried out. In the case $n = 2$ the function $\psi_k$ could have been constructed using the base 5:

$$d_k = \sum_{r=1}^{k} \frac{i_r}{5^r},$$

$i_1 = 0, 1, 2, 3, 4, 5$ and otherwise $i_r = 0, 1, 2, 3, 4$.

There is no conceptual difference between constructions using the decimal base, to using a base $\gamma \geqslant 2n + 1$ for the general case.

Mario Köppel [57] provided an elegant recursive formula for constructing approximating point functions $\psi_k(d_k)$ for any value of $k$. In the process he corrected a numerical error in the original construction in (Sprecher [94, 95]).[3] It is:

$\psi_1(d_1) = d_1$, and for $k > 1$,

$$\psi_k(d_k) = \begin{cases} \psi_{k-1}(d_{k-1}) + \frac{i_k}{10^{\beta_n(k)}}, i_k = 0, 1, ..., 8 & (3.8) \\ \frac{1}{2}\left[\psi_k\left(d_k - \frac{1}{10^k}\right) + \psi_{k-1}\left(d_k + \frac{1}{10^k}\right)\right], i_k = 9 \end{cases}$$

The reader might try to derive from this the formulas for $\psi_3(d_3)$.

The functions $\psi_k(x)$ derived from this definition are such that their limit: $\psi(x) = \lim_{k \to \infty} \psi_k(x)$ is continuous and monotonic increasing. Furthermore, the functions

$$y_q = \psi(x_1 + qa) + \lambda\psi(x_2 + qa)$$

$q = 0, 1, 2, 3, 4$ with constant

$$\lambda = \sum_{r=1}^{\infty} \frac{1}{10^{\beta(r)}} \qquad (3.9)$$

satisfy condition (d) of Theorem 5.

We infer from the definition of $\psi_k$ that

$$\psi_k\left(d_k + \frac{8}{10^{k+1}}\right) = \psi_k(d_k) + \frac{8}{10^{\beta_2(k+1)}},$$

---

[3]See also Jürgen Braun [19].

$$\psi_k \left( d_k + \frac{8}{10^{k+1}} + \frac{8}{10^{k+2}} \right) = \psi_k(d_k) + \frac{8}{10^{\beta_2(k+1)}} + \frac{8}{10^{\beta_2(k+2)}},$$

and in general:

$$\psi_k \left( d_k + 8 \sum_{r=k+1}^{N} \frac{1}{10^r} \right) = \psi_k(d_k) + 8 \sum_{r=k+1}^{N} \frac{1}{10^{\beta(r)}} \qquad (3.10)$$

As $N \to \infty$, therefore:

$$\psi_k \left( d_k + \delta_k \right) = \psi_k(d_k) + 8\varepsilon_k,$$

where

$$\varepsilon_k = \sum_{r=k+1}^{\infty} \frac{1}{10^{\beta(r)}}. \qquad (3.11)$$

Because the function values $\psi_k(d_k)$ are rational, we know that inequality (3.1) always holds, and from the construction we infer that

$$\left| \left[ \psi_k(d_{1,k}) + \lambda \psi_k(d_{2,k}) \right] - \left[ \psi_k(d'_{1,k}) + \lambda \psi_k(d'_{2,k}) \right] \right| \geqslant \frac{1}{10^{2\beta(k)}} \qquad (3.12)$$

It is a matter of calculations to show that

$$8\varepsilon_k < \frac{1}{10^{2\beta(k)}},$$

and we now know that the images of squares $S_k(d_{1,k}, d_{2,k})$ are pairwise disjoint intervals

$$T_k(d_{1,k}, d_{2,k}) = [\psi_k(d_k), \psi_k(d_k) + 8\varepsilon_k] \qquad (3.13)$$

We know now the reason for the choice of numbers $\beta(k)$. They increase fast enough to guarantee that the intervals (3.13) do not intersect for any value of $k$. We conclude that the preceding steps, especially the last two, are true for the limit function $\psi(x)$. A very detailed treatment of these constructions, including carefully worked proofs, can be found in (Braun [19]).

The reader may be able to glean from graphs 3.2, 3.3, and 3.5 that the limit function $\psi(x)$ is not smooth; in fact, it is pathological. This applies equally to Kolmogorov's functions $\psi_{pq}(x)$. Because $\psi(x)$ is strictly monotonic increasing, we know that it has a derivative almost everywhere (meaning everywhere except for a set of measure zero). It can be shown that no derivative exists at the end-points of the intervals $E_k(d_k)$ for all values of $k$. When the derivative does exist, its value is either zero or infinity (Sprecher [87]).

With all that, the behavior of $\psi(x)$ is not too unwieldy; it is subject to a Lipschitz condition $Lip(\log_{10} 2)$:

$$|\psi(x) - \psi(x')| \leqslant c\,|x - x'|^{\log_{10} 2} \qquad (3.14)$$

for a constant $c$ for all points $x \neq x'$ [90, 115].

We shall use a heuristic argument to show this using the prototypical Figure 3.1 and 3.2 illustrating the steps in construction. From these, we deduce that on any interval

$$\left[\frac{i_k}{10^k}, \frac{i_k + 1}{10^k}\right]$$

for $k \geqslant 2$:

$$\left|\psi\left(\frac{i_k + 1}{10^k}\right) - \psi\left(\frac{i_k}{10^k}\right)\right| \leqslant \frac{1}{2^{k-1}}. \qquad (3.15)$$

The Lipschitz condition requires this difference to be related to the length of the interval, and this means finding a constant such that $\frac{1}{2^{k-1}} \leqslant \frac{1}{10^{ak}}$. Taking logarithms of both sides gives

$$a \leqslant \left(1 - \frac{1}{k}\right)\log_{10} 2 < \log_{10} 2.$$

If $x$ and $x'$ are points in that interval, the inequality (3.15) may require a constant $c$ to adjust for the relation between the differences $|\psi(x) - \psi(x')|$ and $|x - x'|$.

An argument based on the self-similarity of the construction of $\psi(x)$ can be used to extend these considerations to another selection of points $x$ and $x'$.

Fridman proved the theorem for the case $n = 2$ with functions $\psi_{pq}(x)$ belonging to class $Lip(1)$:

$$|\psi_{pq}(x) - \psi_{pq}(x')| \leqslant c\,|x - x'|.$$

Whereas Kolmogorov's theorem and all constructions based on it used systems of covering squares set-up in advance, Fridman specified them concurrent with the construction of these functions. Specifically, Kolmogorov's-based intervals maintained a fixed ratio between the diameters of the squares and the widths of the gaps separating them. These ratios determine the behavior of the functions $\psi_{pq}(x)$. In a crucial departure, these ratios were determined by induction, thereby gaining dynamic control over them.

Fridman's approach of an inductive selection of intervals and gaps is still awaiting a constructive version.

With the understanding of the anatomy we acquired, we can show that Kolmogorov's representations are possible with it.

## 3.2   Representations when $n = 2$

The term *Function Representations* is being used interchangeably for two very distinct concepts. The first is mathematical, springing out of abstract ideas that lead to existence theorems – propositions that tell us that these mathematical objects exist, but like oil undersea, does not tell us how realize them. This usage falls under the umbrella of functional complexity whose aim is to find qualitative measures for classification of functions.

The second usage is for numerical constructions aimed at superposition models that can be computed and used for approximations. Mathematical arguments using convergence have a dual role:

They are the method that ensures the existence of these mathematical objects, and provide a basis of techniques for computing with guaranteed increasing accuracy.

Constructions with functions of several variables are mostly carried out through manipulations of symbols, depriving us of an intuitive understanding of shapes and forms and visual renditions. Function representations are presented for this reasons first for functions of two variables. A second reason is that different methods will be presented for representations of functions of more than two variables.

The construction of superpositions and verification that they exist do involve the mathematical concept of *convergence*, that of sequences and series of functions. This is an umbrella term of two different concepts, and it seems appropriate to digress briefly to remind the reader of some basic calculus.

Two types of convergence must be kept apart: *uniform convergence* and *pointwise convergence* . These concepts are not dimension related and we explain them informally in the simplest situation of functions of one variable. Having said this:

Let $\{f_k\}$ be a sequence of functions defined on a domain $A$:

1. The sequence *converges uniformly* on $A$ to a continuous function $f$ if for every error $\varepsilon > 0$ there exists an integer $N$ such that $|f - f_k| < \varepsilon$ for all integers $r > N$;

2. The sequence $\{f_k\}$ *converges pointwise* on $A$ to a function $f$ if $f_k$ converges to $f$ at every point of $A$.

In either case we write $\lim_{k \leftarrow \infty} f_k = f$, and the mode of convergence (a) or (b) must be spelled out.

For series of functions the so-called M-test is available:

Let a sequence of functions $f_k$ be such that $\|f_k\| \leqslant M_k$. If the series $\sum_{r=1}^{k} M_r$ converges, the series $\sum_{r=1}^{k} f_r$ converges uniformly (to a continuous function).

To note the obvious: uniform convergence implies pointwise convergence, but the converse is false. The significant difference between these two modes of convergence is that uniform convergence of a sequence or series of continuous functions guarantees the continuity of the limit function, whereas the pointwise limit of continuous functions does not guarantee that.

This is readily illustrated with the example of the sequence of functions $f_k(x) = x^k$ defined on the unit interval $0 \leqslant x \leqslant 1$. This sequence converges pointwise to the discontinuous function

$$f(x) = \begin{cases} 0 \ at \ points \ 0 \leqslant x \leqslant 1 \\ 1 \ at \ x = 1 \end{cases}.$$

The reason is that $f_k(1) = 1$ for all values of $k$, whereas $\lim_{k \to \infty} f_k(x) = 0$ at any point $x < 1$, as is easily confirmed.

The sequence does not converge uniformly because for any error $\varepsilon > 0$ there is no positive integer $N$ for which $|f(1) - f_k(x)| = 1 - x^k < \varepsilon$ when $n > N$.

Taking this a step further, a function $f(x)$ is continuous at a point $a$, if $\lim_{x \to a} f_k(x) = f(a)$ (of course we have to use left or right limits at the endpoints of E). But also if $\lim_{k \to \infty} f_k(x) = f(x)$ then:

$$\lim_{x \to a} \lim_{k \to \infty} f_k(x) = \lim_{x \to a} f(x)$$

and

$$\lim_{k \to \infty} \lim_{x \to a} f_k(x) = \lim_{k \to \infty} f_k(a) = f(a)$$

and we are led to conclude that these limits are interchangeable:

$$\lim_{k \to \infty} \lim_{x \to a} f_k(x) = \lim_{x \to a} \lim_{k \to \infty} f_k(x).$$

The informed reader will note that this informal digression does not follow completely mathematical orthodoxy; its purpose is merely to remind general readers of facts that hold they key to underlying considerations and choices in many of the specific constructions.

How do these concepts come into play? Both establishing the existence of function representations and approximating a given function are pursued through convergent sequences and series of function. As the above example demonstrates, uniform convergence must be the yardstick if continuity is to be preserved.

We are going now to establish the formula:

$$
\begin{cases}
y_q = \psi(x_1 + qa) + \lambda\psi(x_2 + qa) \\
f(x_1, x_2) = \sum_{q=0}^{4} g_q(y_q)
\end{cases},
$$

with constants $a = \frac{1}{9\cdot10}$ and $\lambda = \sum_{r=1}^{\infty} \frac{1}{10^{\beta(k)}}$; the reader is reminded that $\beta(k) = 2^k - 1$. With families of squares $S_{q,k}(d_{1,k}, d_{2,k})$ defined as in (2.17), we know that for each $q$ and $k$ these squares are mapped onto non-intersecting intervals $T_{q,k}(d_{1,k}, d_{2,k})$.[4]

Remember also that at each point $(x_1, x_2)$ of the unit square, a target function $f(x_1, x_2)$ is being evaluated at the five points $(x_1 + qa, x_2 + qa)$, and so are approximating function $f_k(x_1, x_2)$ to $f(x_1, x_2)$.

We also know that each point $(x_1, x_2)$ lies in at least three squares $S_{q,k}(d_{1,k}, d_{2,k})$, and consequently not less than three of the points $y_q = \psi(x_1 + qa) + \lambda\psi(x_2 + qa)$ lie in intervals $T_{q,k}(d_{1,k}, d_{2,k})$. At these the target function can be approximated to within a predetermined error.

Is all this really sufficient to establish the validity of representations?

To explain the underlying mechanics of the process, we precede the demonstration of validity with a simple heuristic model that

---

[4]See Figure 2.

shows the effectiveness of Kolmogorov's strategy of computing in translated domain.

Even though function representations apply only trivially to functions of one variable, this setting affords a simple graphical demonstration of Kolmogorov's technique. In this case, three summands are approximating a function of one variable.

This is Illustrated this with the simple functions $y = x$ and $y = x^2$ (Figure 3.6 and 3.7). These functions are approximated by discontinuous step functions defined on pairwise disjoint intervals separated by narrow gaps, and then translated by the width of the gaps. The computed values on each of the families are averaged where the intervals overlap; they don't approximate the target function in the gaps.

Observe how even the first iteration of these rough approximations straddle the target functions. Further iterations would interpolate the differences between successive approximations.

These approximations reflect the effectiveness of the strategy of translations. They explain the remarkable degree of approximation of translated superpositions that will be observed later.

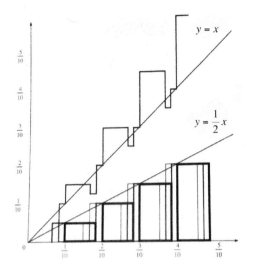

Figure 3.6: First approximation to $y = x$ with two translations.

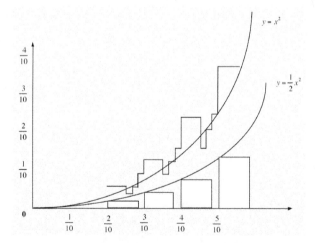

Figure 3.7: First approximation to $y = x^2$ with three translations

Assuming the function $\psi$ in representation given, we shall follow the insightful reasoning of Lorentz [90] in giving an approximation process for constructing the functions $g_q(y_q)$ that compute a target function $f(x_1, x_2)$.

The strategy is to use the squares and intervals defined earlier to construct a series of continuous functions that converged uniformly to $f(x_1, x_2)$. To this end, we pick a positive integer $k = k_1$ and define functions $g_{q,k_r}(y_q)$ having the constant values $f(d_{1,k_1}, d_{2,k_1})$ on the intervals $T_{q,k_1}(d_{1,k_1}, d_{2,k_1})$; that is

$$g_{q,1}(y_q) = \frac{1}{3} f(d_{1,k_1}, d_{2,k_1})$$

when

$$(x_1, x_2) \in T_{q,k_1}(d_{1,k_1}, d_{2,k_1}).$$

for each value of $q$.

We connect the steps with line segments in the gaps as illustrated in Figure 3.8, and designate the continuous function $g_{q,1}(y_q)$.

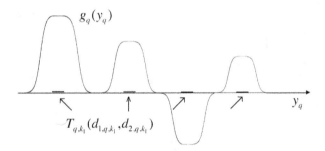

Figure 3.8: The function $g_{q,1}(y_q)$ for fixed $q$.

Note that $g_{q,1}(y_q)$ is a continuous function of one variable along the axis, and $g_{q,1}(y_q) = g_{q,1}[\psi(x_1+qa)+\lambda\psi(x_2+qa)]$ is a continuous function in the plane.

Setting now

$$f_1(x_1, x_2) = \sum_{q=0}^{4} g_{q,1}(y_q) = \sum_{q=0}^{4} g_{q,1}[\psi(x_1 + qa) + \lambda\psi(x_2 + qa)]$$

lets us see how small the difference $\|f - f_1\| < \varepsilon$ can be made with an appropriate choice of $k_1$. The double bars $\|\ \|$ are the customary designation for the maximum value of $f$, on the unit square in our case.

Emulating a routine calculus exercise, pick an error $\varepsilon > 0$, and let the integer $k_1$ be so large that the fluctuation of $f(x_1, x_2)$ on any square $S_{q,k_1}(d_{1,k_1}, d_{2,k_1})$ does not exceed $\varepsilon \|f\|$.[5]

We know that we can do this, because the functions are continuous; by definition, $\|g_{q,1}(y_q)\| < \frac{1}{3}\varepsilon$.

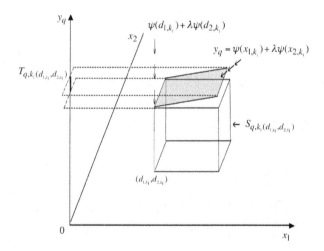

Figure 3.9: Detail of $\sum_{q=0}^{4} g_{q,1}[\psi(x_1 + qa) + \lambda\psi(x_2 + qa)]$

The fact that every point $(x_1, x_2) \in E^2$ belongs to at least three squares $S_{q,k_1}(d_{1,k_1}, d_{2,k_1})$ for three values $q_1, q_2, q_3$ now comes into

---

[5]See Figure 3.9.

play (remember Figure 1). For these values,

$$\|f(x_1, x_2) - g_{q,1}(y_q)\| < \frac{1}{3}\varepsilon \|f\| .$$

Putting all this together we find that

$$\|f(x_1, x_2) - f_1(x_1, x_2)\| = \left\|f(x_1, x_2) - \sum_{q=0}^{4} g_{q,1}(y_q)\right\|$$

$$\leqslant \left\|f(x_1, x_2) - \sum_{r=0}^{3} g_{q_r,1}(y_{q_r})\right\| + \frac{2}{3}\|g_{q,1}(y_q)\|$$

$$\leqslant 3\left\|\frac{1}{3}f(x_1, x_2) - \frac{1}{3}f_1(d_{1,k_1}, d_{2,k_1})\right\| + \frac{2}{3}\|f\|$$

$$\leqslant \varepsilon \|f\| + \frac{2}{3}\|f\| = \left(\varepsilon + \frac{2}{3}\right)\|f\|$$

and we have the first approximating function to $f$.

We went to the length of displaying the spatial variables $(x_1, x_2)$ in these expressions to underscore that the functions $g_{q,1}(y_q)$ are used as functions of two variables in these approximations.

Now iterate this process by applying it to the difference

$$f(x_1, x_2) - f_1(x_1, x_2)$$

instead of $f(x_1, x_2)$: This means selecting a number $k_2 > k_1$ for which the fluctuation of this difference on any square $S_{q,k_2}(d_{1,k_2}, d_{2,k_2})$ does not exceed $\varepsilon \|f - f_1\|$, and repeating the above steps. This begins with the definition

$$g_{q,2}(y_q) = \frac{1}{3}[f(d_{1,k_2}, d_{2,k_2}) - f_1(d_{1,k_2}, d_{2,k_2})]$$

when $(x_1, x_2) \in T_{q,k_2}$, and extending this to a continuous function $g_{q,2}(y_q)$ for which

$$\|g_{q,2}(y_q)\| \leqslant \frac{1}{3}\left(\varepsilon + \frac{2}{3}\right)\|f\| .$$

This is followed with the definition; with

$$f_2(x_1, x_2) = \sum_{q=0}^{4} g_{q,2}(y_q),$$

and a repetition of the preceding routine:

$$\|(f - f_1) - f_2\| \leqslant \left(\varepsilon + \frac{2}{3}\right) \|f - f_1\| \leqslant \left(\varepsilon + \frac{2}{3}\right)^2 \|f\|,$$

and

$$\|f_2\| \leqslant \frac{1}{3} \|f - f_1\| \leqslant \frac{1}{3} \left(\varepsilon + \frac{2}{3}\right) \|f\|.$$

Repeating this process for $(f - f_1 - f_2) - f_3..., ...$ leads to the estimates

$$\left\| f - \sum_{s=0}^{r} f_s \right\| \leqslant \left(\varepsilon + \frac{2}{3}\right)^r \|f\|$$

$$\|g_{q,r}\| \leqslant \frac{1}{3} \left(\varepsilon + \frac{2}{3}\right)^{r-1} \|f\|.$$

To ensure that $\left(\varepsilon + \frac{2}{3}\right)^r \to 0$ as $r \to \infty$, we select $0 < \varepsilon < \frac{1}{3}$. From this we conclude that the series $\sum_{r=1}^{\infty} f_r$ and $\sum_{r=1}^{\infty} g_{q,r}$ converge uniformly to continuous function $f$ and $g_q$ respectively and the representation formula follows.

Figure 3.8 seems to suggest a method for computing the functions $g_q$ through the straightforward approximations described above, but these approximations assume that the intervals

$$T_{q,k_r}(d_{1,k_r}, d_{2,k_r})$$

are given in a linear order for fixed $q$, and this is a fly in this ointment: The function $y = \psi(x_1) + \lambda\psi(x_2)$ (and its translates) scrambles the grid-points $(d_{1,k}, d_{2,k})$, and no simple recursive formula for determining the linear order of the intervals $T_{q,k}(d_{1,k}, d_{2,k})$ is available.

The difficulty in actually constructing the functions $y = \psi(x_1) + \lambda\psi(x_2)$ is a significant one, and it has not been resolved. Nevertheless, computational algorithms have subsequently been developed.[6]

It is worth emphasizing an essential feature of the proof: The two simultaneous approximations to the functions as functions of one variable, and to the functions as functions of two variables. That is, these functions determine simultaneously one-dimensional and two-dimensional graphs, with attendant dual convergence. The above arguments can readily be adapted to any number of variables.

We end this section with an interesting sideways view of Kolmogorov's superpositions that we describe in the case $n = 2$: Consider the function $y = \psi(x_1) + \lambda\psi(x_2)$. Its level-curves:

$$\psi(x_1) + \lambda\psi(x_2) = \text{const.}$$

are monotonic decreasing; each curve begins and terminates on the boundary of $E^2$, and therefore intersects the main diagonal. Therefore also the functions

$$\varphi(x_1, x_2) = g_1[\psi(x_1) + \lambda\psi(x_2)]$$

are constant on these level curves. This is the case with the four translates and functions $g_q$.

Pursuing this line of thought further: the value of each function $g_q$ at any point $(x_1, x_2) \in E^2$ is its value on the intersection

$$[\psi(x_1) + \lambda\psi(x_2)] \cap [x_1 = x_2].$$

This implies that the values of each function $g_q$ are determined on the diagonal $x_1 = x_2$. What this says about the superposition representation of an arbitrary continuous function $f(x_1, x_2)$ is that it

---

[6]See Chapter Four.

is determined at five points on the diagonal through its representation. Compare this with Figure 3.

The first thing to notice is the relation between the level sets of superpositions and target functions. An arbitrary function $f(x_1, x_2)$ may have infinitely many values on one or more level sets $\psi(x_1 + qa) + \lambda\psi(x_2 + qa) = const$. But on these the functions $g_q[\psi(x_1 + qa) + \lambda\psi(x_2 + qa)]$ can compute only one value of $f$.

This is particularly interesting when we think about Kolmogorov's 1956 runner-up paper that relied on Kronrod's discovery that the components of level sets of any function $f(x_1, x_2)$ form a tree. For each of the functions $\psi(x_1 + qa) + \lambda\psi(x_2 + qa)$, however, their level sets are a stick tree; that is, a tree consisting of the diagonal. This points to the intricate ranking of the rational grid points of the unit square.

It is remarkable that this problem was solved with only five summands ($2n+1$ in the general case). It is in these considerations that Figures 3.6 and 3.7 offer an explanation through the strategy of overlapping of translations.

Is five the smallest number of summands for establishing Kolmogorov's function representations of functions of two variables? The proof that we just presented explains how this number ensures convergence to the target function, but we cannot conclude that an alternative proof with fewer summands would not be possible.

That fewer than five summands would not do in the case $n = 2$ was answered by Raouf Doss with geometric arguments using the monotonicity of the functions $\psi_{pq}(x)$ [28]. Let's illustrate this by showing that Kolmogorov's representations cannot be established with only two summands.

In a possible representation

$$f(x_1, x_2) = g_1[\psi_{1,1}(x_1) + \psi_{2,1}(x_2)] + g_2[\psi_{1,2}(x_1) + \psi_{2,2}(x_2)],$$

recall that the curves

$$y_1 = \psi_{1,1}(x_1) + \psi_{2,1}(x_2)$$
$$y_2 = \psi_{1,2}(x_1) + \psi_{2,2}(x_2)$$

are monotonic decreasing with beginning and endpoints on the boundary of $E^2$. Doss demonstrated that there are four points $P_1, P_2, P_3, P_4$ in the square at which two pairs of level curves intersect in a configuration as depicted in Figure 3.10.

If $f(x_1, x_2)$ is a continuous function with values

$$f(P_1) = f(P_2) = 1$$

and

$$f(P_3) = f(P_4) = -1,$$

then

$$g_1(a_1) + g_2(b_1) = 1$$
$$g_1(a_2) + g_2(b_2) = 1$$
$$g_1(a_1) + g_2(b_2) = -1$$
$$g_1(a_2) + g_2(b_1) = -1$$

But adding these pairs gives the contradictory results

$$g_1(a_1) + g_2(b_1) + g_1(a_2) + g_2(b_2) = 2 = -2.$$

The algorithm by which these intersections are obtained can be found in [28]. Chasing down systematically chains of intersections increasing intricacy in the cases of three and four summands, Doss showed that there are continuous functions $f(x_1, x_2)$ not representable with fewer than five summands.

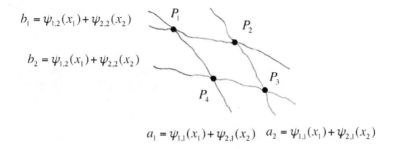

$$b_1 = \psi_{1,2}(x_1) + \psi_{2,2}(x_2)$$

$$b_2 = \psi_{1,2}(x_1) + \psi_{2,2}(x_2)$$

$P_1$   $P_2$

$P_3$

$P_4$

$$a_1 = \psi_{1,1}(x_1) + \psi_{2,1}(x_2) \quad a_2 = \psi_{1,1}(x_1) + \psi_{2,1}(x_2)$$

Figure 3.10: Intersecting level curves according to Doss

Based in part on Doss, Yaki Sternfeld [101] obtained an alternative and considerably simplified proof that five summands is necessary in the case of functions of two variables. In the process he also improved on the result of Bassalygo [9] who proved that three summands are not sufficient.

## 3.3    Representations for arbitrary $n > 2$

Switching from the base 10 to a new base $\gamma \geq 2n+2$ in the computations of the previous two sections brings into focus the intrinsic dependence of the constructions on $n$. This change would allow for the extension of these results to target functions for any integer.

Most notably, families of $n$-dimensional cubes would be needed to cover the cube $2n$ times by translating them along the main diagonal. Substituting a base would satisfy the conditions of Theorem 5: Function Representations.

For the purpose of generalization from two to $n$ variables, we introduce the rational numbers

$$d_{q,k}^{\gamma} = \sum_{r=1}^{k} \frac{i_k}{\gamma^k}$$

for $i_1 = 0, 1, 2, \ldots, \gamma$ and $i_r = 0, 1, 2, \ldots, \gamma-1$ for $r > 1$. Emulating the definition of the intervals $E_{q,k}(d_k)$

$$E_{q,k}(d_k^\gamma) = \left[ d_{q,k}^\gamma - \frac{q}{(\gamma-1)\gamma}, d_{q,k}^\gamma + \delta_k - \frac{q}{(\gamma-1)\gamma} \right] \qquad (3.16)$$

where this time

$$\delta_k = \frac{\gamma-2}{(\gamma-1)\gamma^k} = (\gamma-2) \sum_{r=k+1}^{\infty} \frac{1}{\gamma^r} \qquad (3.17)$$

These intervals are separated by gaps of width $\frac{1}{(\gamma-1)\gamma^k}$. The Cartesian products:

$$S_{q,k}(d_{1,k}, ..., d_{n,k}) = \prod_{p=1}^{n} E_{q,k}(d_{p,k}^\gamma),$$

and the new setting do not involve conceptual changes to what we have done in the previous section.

The function $\psi(x)$ and consequently the devil's-staircase functions $\psi_k$ used to define it require redefinition, using the base $\gamma$. Mario Köppen's recursion formula (3.8) becomes for $k > 1$:

$$\psi_k(d_k^\gamma) = \begin{cases} \psi_{k-1}(d_{k-1}^\gamma) + \frac{i_k}{\gamma^{\beta_n(k)}}, & i_k = 0, 1, ..., \gamma - 2 \\ \frac{1}{2}\left[ \psi_k\left(d_k^\gamma - \frac{1}{\gamma^k}\right) + \psi_{k-1}\left(d_k^\gamma + \frac{1}{\gamma^k}\right) \right], & i_k = \gamma - 1 \end{cases} \qquad (3.18)$$

we already encountered the versions:

$$\begin{cases} y_q = \sum_{p=1}^{n} \lambda_p \psi(x_p + qa) \\ f(x_1, ..., x_n) = \sum_{q=0}^{2n} g_q(y_q) \end{cases} \qquad (3.19)$$

for which we select the numbers $\lambda_1 = 1$ and

$$\lambda_p = \sum_{r=1}^{\infty} \frac{1}{\gamma^{(p-1)\beta_n(r)}}, p > 1,$$

where

$$\beta_n(k) = \frac{n^k - 1}{n - 1} = 1 + n + n^2 + \dots + n^{k-1};$$

the constant $a$ will be defined subsequently.

This is a direct generalization of representations for the case $n = 2$, and all arguments and calculations follow verbatim. Pointing out the obvious again: with base $\gamma \geqslant 2n + 1$, all continuous functions of $2 \leqslant m \leqslant n$ variables can be represented. All you need to do is to start with $k = m$ when representing a function of $m$ variables.[7]

By now we have meandered through many computations establishing different aspects of superpositions, and a quick review of the essential role of the numbers $\beta_n(k)$ appears to be called for. This is particularly cogent because these calculations were carried out for the case $n = 2$.

In terms of $\gamma$, the image intervals of cubes $S_{q,k}$ are determined by the functions

$$y_q = \sum_{p=1}^{n} \lambda_p \psi(x_p + qa).$$

Their rate of increase is determined by the rate of increase of $\psi(x)$, and this brings us to the numbers $\beta_n(k)$: You recall from the construction that they determined the maximum rate $\frac{\gamma - 2}{\gamma^{\beta_n(k)}}$ at which $\psi(x)$ increases on the intervals $E_{q,k}(d_k^\gamma)$.

Without loss of generality let's focus on $q = 0$:

$$\psi_k(d_k^\gamma + \delta_k) = \psi_k(d_k^\gamma) + (\gamma - 2)\varepsilon_k \tag{3.20}$$

where[8]

$$\varepsilon_k = \sum_{r=k+1}^{\infty} \frac{1}{\gamma^{\beta_n(r)}}.$$

---

[7]See in this connection Brattka [18] and Lin-Unbehauen [63].
[8]See (3.11).

Accordingly,

$$\sum_{p=1}^{n} \lambda_p \psi(x_p + \delta_k) = \sum_{p=1}^{n} \lambda_p \psi(x_p) + (\gamma - 2)\varepsilon_k \sum_{p=1}^{n} \lambda_p,$$

and the image intervals of the cubes $S_k$ are

$$T_k(d_{1,k}, \ldots, d_{n,k}) = \left[ \sum_{p=1}^{n} \lambda_p \psi(x_p), \sum_{p=1}^{n} \lambda_p \psi(x_p) + (\gamma - 2)\varepsilon_k \sum_{p=1}^{n} \lambda_p \right].$$

$$(3.21)$$

This shows that the numbers $\beta_n(k)$ determine the widths of these intervals to ensure their disjointness, thereby meeting requirement (d) in Theorem 5.

These considerations reveal where the number $n$ comes into play. Specifically, (3.18) shows that $n$ is an intrinsic parameter in the construction of the function $\psi$. The rational numbers on which all functions are sampled also depend on $n$, as do the functions $g_q$ that compute $f$, but this, of course, is expected.

The number $n$ as a parameter cannot, of course, be eliminated entirely from function representations with variable-reducing superpositions. The question arises about the possibility of shifting $n$ to the implementation stage, confining its presence to the direct computations of $f$. We ask if there are a priori constructions that are independent of $n$ for representing every function of $n \geqslant 2$ variables? Since such a priori constructions are rooted in $\psi(x)$, the question is if $n$ can be eliminated as a parameter in constructing $\psi(x)$?

This dependence of $\psi$ on $n$ has particular relevance if we want to represent continuous functions of any number of variables efficiently, and if we think in terms of a computer that stores the function once and for all.

There is an alternative strategy for constructing a universal continuous function $\tilde{\psi}(x)$ that is independent of $n$ and can be used in the representation of any continuous function $f(x_1, \ldots, x_n)$ for any $n \geq 2$. This strategy gives Theorem 6.[9]

**Theorem 6.** *There exists a continuous function $\tilde{\psi}(x)$ such that every continuous function $f(x_1, \ldots, x_n)$ defined on the unit cube $E^n$ with $n \geqslant 2$ has a representation*

$$f(x_1, \ldots, x_n) = \sum_{q=0}^{2n} g_q \circ \sum_{p=1}^{n} \lambda_p \tilde{\psi}(x_p + qa) \tag{3.22}$$

*with continuous functions $g_q$ and suitable constants $a$ and $\lambda_p$.*

Beginning with an a priory selection of an infinite sequence of decreasing rationally independent transcendental numbers $\lambda_r$, $\sum_{r=1}^{\infty} \lambda_r$, the universal function $\tilde{\psi}(x)$ will be constructed on the interval $D_1 = \left[0, 1 + \frac{1}{5!}\right]$ with rational numbers $\rho_1 > \rho_2 > \rho_3 > \ldots$ determined through a bootstrapping procedure.

The beginning value $5!$ of the constructions that follow was chosen because lower values are not useful in implementing formula (3.22).

Like the function $\psi(x)$, the function $\tilde{\psi}(x)$ is obtained as the limit of a sequence of point-functions $\tilde{\psi}(\tilde{d}_k)$ that are extended to piecewise continuous functions. The rational numbers $\tilde{d}_k$ use a factorial base:

$$\tilde{d}_k = \sum_{r=1}^{k} \frac{i_r}{(r+4)!} \tag{3.23}$$

$i_1 = 0, 1, \ldots, 5! + 1$, $i_r = 0, 1, \ldots, r - 1$ otherwise.

The graph of $\tilde{\psi}_1$ is defined through point on the main diagonal:

$$\tilde{\psi}_1\left(\frac{i_1}{5!}\right) = \frac{i_1}{5!}, \quad i_1 = 0, 1, \ldots, 5! + 1.$$

---

[9]See Sprecher 92, 93.

Selecting a rational number $\rho_1 < \frac{1}{6!}$, we define $\tilde{\psi}_2\left(\frac{i_2}{6!}\right)$ on the interval $D_0 = \left[0, \frac{1}{5!}\right]$ as follows:

$$\tilde{\psi}_2\left(\frac{i_2}{6!}\right) = \begin{cases} \frac{i_2}{5}\rho_1, & i_2 = 0, 1, 2, 3, 4 \\ \frac{1}{2}\left[\frac{4}{5}\rho_1 + \frac{1}{5!}\right], & i_2 = 5 \end{cases}$$

$$\tilde{\psi}_2\left(\frac{1}{5!}\right) = \frac{1}{5!}.$$

The graph of this function is similar to Figure 3.2 with a change of scales. On the interval $D_1$ we define

$$\tilde{\psi}_2\left(\frac{i_1}{5!} + \frac{i_2}{6!}\right) = \frac{i_1}{5!} + \begin{cases} \frac{i_2}{5}\rho_1, & i_2 = 0, 1, 2, 3, 4 \\ \frac{1}{2}\left[\frac{4}{5}\rho_1 + \frac{1}{5!}\right], & i_2 = 5 \end{cases}$$

Let

$$\rho_2 < \mu_2 = \frac{1}{\lambda_1 + \lambda_2} \min_{\tilde{d}_p, \tilde{d}'_p} \left| \sum_{p=1}^{2} \lambda_p [\tilde{\psi}_p(\tilde{d}_p) - \tilde{\psi}_p(\tilde{d}'_p)] \right|$$

and define the point-function $\tilde{\psi}_3\left(\frac{i_3}{7!}\right)$ on the interval $[0, 1/6!]$:

$$\tilde{\psi}_3\left(\frac{i_3}{7!}\right) = \begin{cases} \frac{i_3}{6}\rho_2, & i_3 = 0, 1, ..., 5 \\ \frac{1}{2}\left[\frac{5}{6}\rho_2 + \frac{1}{5}\rho_1\right], & i_3 = 6 \end{cases}$$

$$\tilde{\psi}_3\left(\frac{1}{6!}\right) = \frac{1}{5}\rho_1$$

This graph is similar to Figure 3.2 with appropriate changes in the scales. The extension to the interval $D_0$ is:[10]

$$\tilde{\psi}_3\left(\frac{i_2}{6!}+\frac{i_3}{7!}\right) = \begin{cases} \frac{i_1}{5}\rho_1 + \begin{cases} \frac{i_2}{6}, & i_2 = 0,1,...,5 \\ \frac{1}{2}\left[\frac{5}{6}\rho_2 + \frac{1}{5}\rho_1\right], & i_3 = 6 \end{cases} & i_2 = 0,1,...,4 \\ \frac{1}{2}\left[\frac{4}{5}\rho_1 + \frac{5}{6}\rho_2 + \frac{1}{2}\left(\frac{4}{5}\rho_1 + \frac{1}{5!}\right)\right], & i_2 = 4 \ i_3 = 6 \\ \frac{1}{2}\left[\frac{4}{5}\rho_1 + \frac{1}{5!}\right], & i_2 = 4 \ i_3 = 5 \\ \frac{1}{2}\left[\frac{5}{6}\rho_2 + \frac{1}{2}\left(\frac{4}{5}\rho_1 + \frac{1}{5!}\right) + \frac{1}{5!}\right], & i_2 = 5 \ i_3 = 6 \end{cases}$$

$$\tilde{\psi}_3\left(\frac{1}{5!}\right) = \frac{1}{5!}$$

The graph of this function can also be understood by relating it to Figure 3.5; the complete function is

$$\tilde{\psi}_3\left(\tilde{d}_3\right) = \tilde{\psi}_3\left(\frac{i_1}{5!}\right) + \tilde{\psi}_3\left(\frac{i_2}{6!}+\frac{i_3}{7!}\right).$$

Assuming the rational number $\rho_{k-1} < \mu_{k-1}$ already determined, we select

$$\rho_k < \mu_k = \frac{1}{\sigma_k}\min_{\tilde{d}_p,\tilde{d}'_p}\left|\sum_{p=1}^{k}\lambda_p[\tilde{\psi}_p(\tilde{d}_p) - \tilde{\psi}_p(\tilde{d}'_p)]\right|,$$

where

$$\sigma_k = \sum_{r=1}^{k}\lambda_r. \tag{3.24}$$

---

[10]See Figure 3.3.

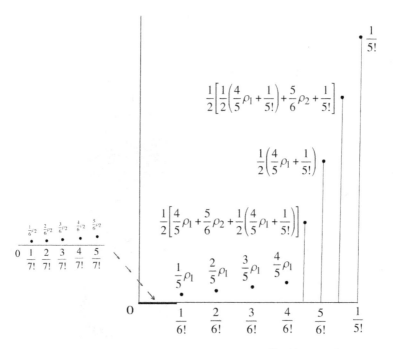

Figure 3.11: The point function $\tilde{\psi}_3 \left( \frac{i_2}{6!} + \frac{i_3}{7!} \right)$

Because the values $\tilde{\psi}_p \left( \tilde{d}_p \right)$ are rational, we know that $\mu_k \neq 0$ when $\sum\limits_{p=1}^{k} \left| \tilde{d}_p - \tilde{d}'_p \right| \neq 0$ for all values of $k$.

The function $\tilde{\psi}_k \left( \tilde{d}_k \right)$ can be derived from the recursion formula (3.18):

$\tilde{\psi}_1 \left( \tilde{d}_1 \right) = \frac{i_1}{5!}$, and for $k > 1$

$$\tilde{\psi}_k \left( \tilde{d}_k \right) = \begin{cases} \tilde{\psi}_{k-1} \left( \tilde{d}_{k-1} \right) + \frac{i_k}{k-1}\rho_k, & i_k = 0, 1, ..., k-2 \\ \frac{1}{2} \left[ \tilde{\psi}_k \left( \tilde{d}_k - \frac{1}{k!} \right) + \tilde{\psi}_{k-1} \left( \tilde{d}_k + \frac{1}{k!} \right) \right], & i_3 = k-1 \end{cases}$$

$$(3.25)$$

The transition to the limit function $\tilde{\psi}(x)$ is as straight forward as the generation of the $n$-dependent $\psi(x)$ through formula 2.18. The above construction is only moderately more complicated than that of constructing $\psi(x)$, but there is a price to pay for eliminating $n$ as a parameter from $\tilde{\psi}(x)$. This becomes evident when we apply it to Theorem 6 for an arbitrary $n \geqslant 2$.

Recalling Theorem 5, given an integer $n \geqslant 2$ we must now construct families of cubes (squares in the case $n = 2$) with the covering property specified there, and concurrently we must show the disjointness of the images of their image intervals. Here the simplicity of uniform intervals with a fixed ratio $(\gamma - 2) : 1$ between their lengths and the widths of the gaps separating them is abandoned.

The fixed ratios are replaced by variable ratios in the new strategy; specifically, the length of intervals exceeds the length of the gaps by a factor of $(k - 1) : 1$. This is conveniently accomplished with intervals

$$\tilde{E}_k(\tilde{d}_k) = \left[ \tilde{d}_k, \tilde{d}_k + \frac{1}{k!} - \sum_{r=k+1}^{\infty} \frac{1}{r!} \right] \tag{3.26}$$

These intervals are clearly separated by gaps of widths

$$\varepsilon_k = \sum_{r=k+1}^{\infty} \frac{1}{r!}.$$

The inclusion property

$$\tilde{E}_k(\tilde{d}_k) \supset \tilde{E}_{k+1}(\tilde{d}_{k+1})$$

when $i_{k+1} = (k+1)i_k + t$ and $t = 0, 1, ..., k - 1$ is verified with a direct calculation.

It follows from the construction that setting

$$\nu_k = \sum_{r=k}^{\infty} \frac{r-1}{r} \rho_r, \tag{3.27}$$

then

$$\tilde{\psi}\left(\tilde{d}_k + \delta_k\right) = \tilde{\psi}\left(\tilde{d}_k\right) + \nu_k. \tag{3.28}$$

If $n \geqslant 2$ is a given integer, and

$$S_k(\tilde{d}_1, ..., \tilde{d}_n) = \prod_{p=1}^{n} E_k(\tilde{d}_{p,k})$$

then their image intervals under the mapping $\sum\limits_{p=1}^{n} \lambda_p \tilde{\psi}_p\left(\tilde{d}_p\right)$ are

$$T_k(\tilde{d}_1, ..., \tilde{d}_n) = \left[\sum_{p=1}^{n} \lambda_p \tilde{\psi}_p\left(\tilde{d}_p\right), \sum_{p=1}^{n} \lambda_p \tilde{\psi}_p\left(\tilde{d}_p\right) + \sigma_k \nu_k\right]. \tag{3.29}$$

To show that these intervals are disjoint for fixed $k$, we need only show that their length is less than $\sigma_k \rho_k$; namely, that $\sigma_k \nu_k < \sigma_k \rho_k$. This is verified with a direct calculation. The highlight of the verification of (3.28) is

$$\tilde{\psi}_k\left(\tilde{d}_k + \sum_{r=k+1}^{N} \frac{r-1}{r}\right) = \tilde{\psi}_k\left(\tilde{d}_k\right) + \sum_{r=k+1}^{N} \frac{r-2}{r-1}\rho_r,$$

$$\sum_{r=k+1}^{N} \frac{r-2}{r-1}\rho_r = \sum_{r=k}^{N-1} \frac{r-1}{r}\rho_r,$$

and consequently

$$\lim_{N\to\infty} \sum_{r=k+1}^{N} \frac{r-2}{r!} = \frac{1}{k!} - \lim_{N\to\infty} \left(\sum_{r=k+1}^{N} \frac{1}{r!} - \frac{1}{N!}\right) = \delta_k.$$

Finally, to establish Theorem 6 for an integer $n \geqslant 2$, select a factorial $k_0! \geqslant 2n + 1$, and insert into representation (3.22)

$$a = \sum_{r=k_0}^{\infty} \frac{1}{r!}.$$

This allows for $2n$ translations as required, and it goes without saying that the above conclusions are true for the translated functions and cubes.

An alternative two-stage construction of an equivalent universal function $\tilde{\psi}_k(x)$ was published in (Sprecher [123]). In that construction, the numbers were determined in a second stage, after the functions $\tilde{\psi}_k\left(\tilde{d}_k\right)$ were constructed with unspecified rational numbers $\rho_k$.

The question about the least number of summands at the close of the previous section is even more relevant here. Understanding why the geometric method in the case $n = 2$ fails when applied to representations with five summands might have illuminated Kolmogorov's superpositions beyond its intent, but Doss pointed out that the method was already strained in dealing with four summands.

This geometric strategy is not amenable to function presentations of more than two variables. Sternfeld pointed out correctly that this aspect of Kolmogorov's function representations is topological rather than analytic. This broadens these concepts beyond their setting in compact Euclidean spaces (in the sense of closed and bounded). We shall touch on these generalizations in the next chapter, and only note here that the number $2n + 1$ of summands cannot be reduced.[11]

---

[11] Sternfeld [99, 100] for a broad dimension-theoretic discussion.

# Chapter 4

# Functional Complexity

## 4.1 An anatomy of superpositions

Returning to Kolmogorov's original formulation of function representations,

$$f(x_1, ..., x_n) = \sum_{q=1}^{2n+1} g_q[\psi_{1,q}(x_1) + \psi_{2,q}(x_2) + ... + \psi_{n,q}(x_n)]$$

we have shown that the functions $\psi_{pq}(x)$ can be replaced by functions of the form $\lambda_p \psi(x + qa)$. Now we show that they cannot be replaced by functions of the form $a_{pq}\psi_p(x)$.

In fact, for any given integer $m$, there is a polynomial that cannot be represented in the form (Sprecher [85])

$$P(x_1, x_2) = \sum_{q=1}^{m} g_q[a_{1,q}\psi_1(x_1) + a_{2,q}\psi_2(x_2)]. \qquad (4.1)$$

Like the functions $\psi_{pq}(x)$, the functions $\psi_p(x)$ are taken to be monotonic increasing, and they have therefore inverses $\psi_p^{-1}(x)$.

Setting $t_p = \psi_p(x_p)$ and substituting $x_p = \psi_p^{-1}(t_p)$ in equation (4.1), then a routine calculus exercise shows that

$$P(\psi_1^{-1}(t_1), \psi_2^{-1}(t_2)) = \sum_{q=1}^{m} g_q[a_{1,q}t_1 + a_{2,q}t_2 + b_q] = U(t_1, t_2)$$

It was observed by Avron Douglis that any continuous function $U(t_1, t_2)$ represented this way would have to satisfy a weak partial differential equation with constant coefficients.

A new concept enters our vocabulary: *weak derivative s*. The quickest way to explain what this means is to sketch the generalization of Solomon Bochner of the concept of derivative (Bochner [16]), but first some basic calculus.

The partial differential operators $\frac{\partial U}{\partial t_1}$ and $\frac{\partial U}{\partial t_2}$ are defined for a differentiable function $U$ as the limits of sequences

$$\begin{aligned} \Delta_1 U &= \frac{U(t_1+\varepsilon, t_2) - U(t_1, t_2)}{\varepsilon} \\ \Delta_2 U &= \frac{U(t_1, t_2+\varepsilon) - U(t_1, t_2)}{\varepsilon} \end{aligned}, \quad \varepsilon > 0$$

as $\varepsilon \to 0$. According to Bochner's theory, these sequences can be given a meaning even when the function $U$ is only continuous. They are said to represent equivalence classes of weakly convergent sequence s, but we don't have to go into that by simply agreeing that the partial differential operators $\frac{\partial U}{\partial t_1}$ and $\frac{\partial U}{\partial t_2}$ are defined through them in that case.

From here on the argument follows that of a standard calculus exercise. Thus, applying the partial differential operator

$$a_{2,q}\frac{\partial}{\partial t_1} - a_{1,q}\frac{\partial}{\partial t_2}$$

for a given value of $q$ to the function

$$h_q = \sum_{p=1}^{2} a_{pq}t_p + b_q$$

gives the following:

$$\left( a_{2,q} \frac{\partial}{\partial t_1} - a_{1,q} \frac{\partial}{\partial t_2} \right) g_q(h_q) =$$

$$a_{2,q} \frac{g_q(h_q + a_{1,q}\varepsilon) - g_q(h_q)}{\varepsilon_1} - a_{1,q} \frac{g_q(h_q + a_{2,q}\varepsilon) - g_q(h_q)}{\varepsilon_1}$$

The substitution $\varepsilon_1 = \frac{\varepsilon'}{a_{1,q}}$ and $\varepsilon_2 = \frac{\varepsilon'}{a_{2,q}}$ shows that each of the difference quotients in the right side equals

$$a_{1,q} a_{2,q} \frac{g_q(h_q + \varepsilon') - g_q(h_q)}{\varepsilon'},$$

and consequently

$$\left( a_{2,q} \frac{\partial}{\partial t_1} - a_{1,q} \frac{\partial}{\partial t_2} \right) g_q(h_q) = 0.$$

Being true for every value of $q$, it follows that

$$\prod_{q=1}^{m} \left( a_{2,q} \frac{\partial}{\partial t_1} - a_{1,q} \frac{\partial}{\partial t_2} \right) U = \prod_{q=1}^{m} \left( a_{2,q} \frac{\partial}{\partial t_1} - a_{1,q} \frac{\partial}{\partial t_2} \right) \sum_{q=1}^{m} g_q(h_q) = 0.$$

This is not satisfied when $U(t_1, t_2) = t_1^{m+1} + \nu(t_2)$ and any polynomial $\nu(t_2)$.

Along these lines, George Pólya and Gábor Szegö have shown that the function $f(x_1, x_2, x_3) = x_1 x_2 + x_1 x_3 + x_2 x_3$ cannot be represented with superpositions of three infinitely differentiable functions of two variables (Pólya and Szägo [79]). This gets us into the area of superpositions subject to smoothness conditions beyond continuity.

These questions are closely associated with the second part of Hilbert's Problem 13, that Kolmogorov's function representations did not address. He nevertheless was interested in these problems, and he made the following specific assertion:

> There exists analytical function of three variables that could not be represented by any finite superposition of continuously differentiable function of two variables. There exists analytical function of two variables that is not represented by any finite superposition of continuously differentiable functions of one variable and addition. (Kolmogorov [56])

Kolmogorov himself did not publish more on this subject, but the impressive school that he built made seminal contributions to it, notably Vituškin and his student Henkin.

Within the realm of continuity in which Kolmogorov's function representations were set, there was no apparent price paid for the representations with functions $\psi_{pq}(x)$ or $\psi(x+qa)$. But there actually was a price, though unspecified. A hint of this can be gleaned from an examination of Figures 4.5 and 4.6 below and the constructions in the representation section for $n = 2$. These suggest that the fluctuations of the functions $g_q(y_q)$ would in general be considerably greater than that of the target function $f$, making them possibly less well behaved. No measure has been developed to make this more than a vague notion.

While all we can say about the superposition is that they are continuous, the functions that they represent include every smoothness class; they can be analytic or have any number of partial derivatives.

Is there anything that can be said about representations of differentiable functions? Can the smoothness of a target function influence the behavior of representing superpositions in a measurable way? More generally, can smooth functions be represented with smooth superpositions?

We already know that not all analytic functions of three variables can be represented with superpositions of analytic functions of two variables, but can more general conclusions be drawn? For example, can functions of $n$-variables with $m$ continuous partial derivatives be represented with superpositions of functions of $p < n$ variables having a given number of continuous partial derivatives?

It turned out that framing function representations with these questions, Hilbert's idea was sound; he just picked the wrong measure. An indicator that combined the number of variables with a number of derivatives was a valid index of classification.

In 1954, three years before the Arnold-Kolmogorov refutation of Hilbert's conjecture, Antoli Georgievich Vituškin proved a seminal result that to date can still be regarded as the deepest insight with these questions. In general terms, he established that there was an inevitable decent in in smoothness of superposition representations with fewer variables, in the sense that smoothness must decrease with a decrease in the number of variables.

Vituškin proved the following (Vituškin [145]):

**Theorem 7.** *Let the functions $f(x_1, ..., x_n)$ have $p \geqslant 1$ continuous partial derivatives, and let the last of these be subject to a Lipschitz condition $Lip(\alpha)$ for a constant $0 < \alpha < 1$. Than not all functions with characteristic $\frac{p+\alpha}{n}$ can be represented with superpositions with characteristic $\frac{p'+\alpha'}{n'} > \frac{p+\alpha}{n}$.*

With Kolmogorov's function representations this means that if a function $f$ with characteristic $\frac{p+\alpha}{n}$ is to be represented with superpositions of characteristic $\frac{p'+\alpha'}{1}$, then the inequality $\frac{p+\alpha}{n} > p'+\alpha'$ must be satisfied. This tells us how much the differentiability of superpositions must decline.

So, for example, not all functions of three variables with three continuous partial derivatives can be represented with three times continuously differentiable functions of two variables.

It is rather astonishing to realize how unexpected Kolmogorov's discovery was as late as three years before its publication, even within his own circle of students. Referring to the 1956 paper that first demonstrated that the number $n$ of variables fails as an index for the classification of continuous functions, Vituškin commented:

Kolmogorov's theorem on the possibility of representing continuous functions of n variables as superposition
of continuous functions of three variables was highly
unexpected.

Vituškin was a student of Kolmogorov who graduated in 1954,
the year he published Theorem 2. Recalling Kolmogorov's own surprise at this breakthrough, it is not surprising that this possibility
hit like a bolt of lightening also in his inner circle.

In an undated lecture,[1] probably given in 1968, Vituškin an-
ticipated positive results on Hilbert's conjecture from a different
direction:

All the results known up until now don't contradict,
for example, the hypothesis that the function $f(x, y, z)$
defined by the 7th degree normalized polynomial equation is not a finite superposition of analytical functions
of two variables.

This was in line also with Kolmogorov's own reading of Problem
13 in its full context that went beyond mere continuity.

A number of other important results is this general direction
have been obtained by Vituškin and Henkin.[2] Their methods and
direction are far removed from the purpose of this monograph and
these results are therefore only mentioned in passing.

Preceding Vituškin's discovery over three decades earlier, Ostrovski's showed in 1920 that the analytic function

$$\zeta(x, y) = \sum_{n=1}^{\infty} \frac{x^n}{n^y}$$

cannot be represented with analytic functions of one variable, and
algebraic functions of any number of variables (Ostrovski [77]).

---

[1]See End Matter and also [109]

[2]See, for example, [57] and [108].

Before turning our Kaleidoscope back to the main path of function presentations, I present one more result of Vituškin that is closely connected to Kolmogorov's theorem. It concerns representations

$$\sum_{n=1}^{N} \varphi_q(x_1, x_2) \cdot g_q[\psi_q(x_1, x_2)]. \tag{4.2}$$

with function-coefficients. For orientation, compare these summands with summands that you are familiar with:

$$\varphi_q(x_1, x_2) \cdot g_q[\psi_q(x_1, x_2)]$$
$$g_q[\psi_q(x_1, x_2)]$$
$$g_q[\psi_{1,q}(x_1) + \psi_{2,q}(x_2)]$$
$$g_q[\psi(x_1) + \lambda\psi(x_2)]$$

In equation (4.2) the functions $\varphi_q$ are arbitrary, the functions $\psi_q$ have continuous partial derivatives, and the functions $g_q$ are continuous. The significant change in format here is in the added presence of the function-coefficients $\varphi_q$. Vituškin has shown that not all analytic functions of two variables have this representation. Henkin, among much work in this area, showed that there is a polynomial $(x_1 + ax_2)^N$ not representable in the form (4.2).

The beauty of Kolmogorov's function representations is their simplicity, using for their establishment only direct computations of complicated but elementary means. Their phrasing circumvented concepts that are by no means as simple, such as the meaning of dimension. This concept has an intuitive meaning when we think of first, second and third dimensions; beyond that we accept the standard extension taught in calculus by adding variables to the familiar equations.

Hidden in Kolmogorov's theorem is a mathematical definition of the concept of *covering dimension*, formulated by Henri Lebesgue.

We saw that the unit square can be covered with three families of disjoint squares; the $n$-dimensional cube can be covered with $n+1$ disjoint cubes. This is how Lebesgue defined dimension, but with a caveat: The covering sets are taken to be open, without boundaries to state this as simple as possible.

You the reader can convince yourself that a circle can be covered with overlapping arcs in such a way that no point belongs to more than two arcs. In general you might think of a space as having dimension $n$ if it can be covered with $n+1$ disjoint sets. From a mathematical point of view this is too vague and imprecise a definition, but it will do for our purposes here.

The Euclidean space setting for Kolmogorov's function representations was a natural one because of their roots in a specific algebraic problem. Just as natural to mathematicians was the question: Are such representations possible under more general settings? The first publication removing the confines of Euclidean space was a ground breaking four-page note titled *Dimension of metric space s and Hilbert's problem 13* by Ostrand [76].

Instead of functions $\psi_{pq}(x)$ defined on the interval $E = [0, 1]$, Ostrand had them defined on a compact (closed and bounded in our context) metric space of finite dimension $X^p$, and $f(x_1, ..., x_n)$ is continuous and defined on the product metric space $\prod_{p=1}^{n} X^p$.

With the exception of these changes, the statement of the representation theorem looks like that of Kolmogorov. But in the establishment of this generalization, arguments use a different language, and we do not pursue this further. This paper is included because of its importance in changing the underlying space of function presentations, and because the next generalization that we mention is based on it.

The choice of the unit cube $E^n$ in phrasing function representations was a convenience to guarantee uniform convergence. This precludes functions such as

$$f(x_1, x_2) = \frac{1}{(x_1 - 1)^2 + (x_2 - 1)^2} \quad (4.3)$$

that is continuous on the open square and not on the coordinate axes.

Reverting to Kolmogorov's function representation original version, Raouf Doss demonstrated that with a total of 8 continuous functions $\varphi_q(x_1, x_2)$ this function has a representation

$$f(x_1, x_2) = \sum_{q=1}^{5} g_1[\varphi_q(x_1, x_2)] + \sum_{q=6}^{8} g_2[\varphi_q(x_1, x_2)]$$

on the open square.

More generally, he showed that continuous functions $f(x_1, ..., x_n)$ defined on the entire space $R^n$ has a representation (Doss [30])

$$f(x_1, ..., x_n) = \sum_{q=1}^{2n+1} g_1[\varphi_q(x_1, ..., x_n)] + \sum_{q=2n+2}^{4n} g_2[\varphi_q(x_1, ..., x_n)].$$

$$(4.4)$$

In both of these formulas, the functions $\varphi_q$ in the first and second sum are continuous but they don't share the same characteristics, as do the continuous functions $g_1$ and $g_2$.

Just as the discovery of the functions $\psi_{pq}(x)$ was the key to Kolmogorov's representation formula, the discovery of appropriate functions $\varphi_q(x_1, ..., x_n)$ was the key to these representations. You will see shortly that the technique used to establish representation (3.4) precluded expressing these with sums of functions$\psi_{pq}(x)$. Look at it as a generalization of Theorem 2.1.

For clarity we look at the case of functions $f(x_1, x_2)$ of two variables defined on the open disk $D^2 = \{x_1, x_2 : x_1^2 + x_2^2 < 1\}$ as shown in Figure 4.1. This allows for unbounded functions, such as (3.3) that tends to infinity as points $(x_1, x_2)$ approach the boundary $x_1^2 + x_2^2 = 1$.

Doss utilizes a system of concentric circles $C_1, C_2, C_3, \ldots$ centered at 0, and with increasing radii $\alpha_1, \alpha_2, \alpha_3, \ldots$ converging to 1 as $r$ tends to infinity.[3]

Eight continuous functions on $D^2$ are constructed in two groups: The functions $\varphi_1, \varphi_2, \varphi_3, \varphi_4, \varphi_5$ are such that

$\varphi_q(x_1, x_2) = \alpha_r$ for $q = 1, 2, 3, 4, 5$ and all values of $r$.

If $F(x_1, x_2)$ is a continuous function on the ring $\alpha_r^2 \leqslant x_1^2 + x_2^2 \leqslant \alpha_{r+1}^2$ such that $F(x_1, x_2) = 0$ on the circles $x_1^2 + x_2^2 = \alpha_r^2$ and $x_1^2 + x_2^2 = \alpha_{r+1}^2$, then it has a representation

$$F(x_1, x_2) = \sum_{q=1}^{5} g_1[\varphi_q(x_1, x_2)]$$

with a continuous function $g_1$. It follows that

$$g_1(\alpha_r) = g_1(\alpha_{r+1}) = 0.$$

Finding the function $g_1$ can be looked at as a boundary-value type problem. You might think of $F(x_1, x_2)$ as being composed of functions defined on rings $\alpha_r^2 \leqslant x_1^2 + x_2^2 = \alpha_r^2$, and vanishing on the boundary $C_r \cup C_{r+1}$ for $r = 1, 2, 3, \ldots$.

On each ring the functions $\varphi_1, \varphi_2, \varphi_3, \varphi_4, \varphi_5$ are constructed through an approximation that is based on coverings like those used in Figure 1. In fact, the intersections of such families with the ring would serve the purpose, and convergence arguments follow the lines of reasoning that we used above.

For the remaining functions $\varphi_6, \varphi_7, \varphi_8$, it is shown that there is a continuous function $g_2$ such that for every continuous function $f(x_1, x_2)$ on $D^2$:

$$f(x_1, x_2) = \sum_{q=6}^{8} g_2[\varphi_q(x_1, x_2)] \text{ when } x_1^2 + x_2^2 = \alpha_r^2.$$

---

[3]See Figure 4.1.

In conclusion, it is shown that at all points $(x_1, x_2) \in D^2$,

$$f(x_1, x_2) = \sum_{q=1}^{5} g_1[\varphi_q(x_1, x_2)] + \sum_{q=6}^{8} g_2[\varphi_q(x_1, x_2)].$$

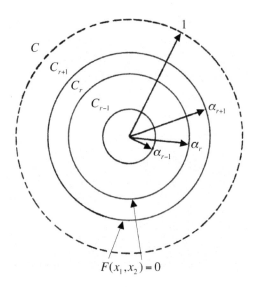

Figure 4.1: Circles $C_{r-1}, C_r, C_{r+1}$ converging to C

We glossed over the detailed arguments that can be found in [37]. Using powerful tool of functional analysis, Jean Pierre Kahane provided the most general proof of Kolmogorov's function representations [65]. In mathematical terms his statement of the theorem is the more elegant and simplest because of its lean conditions, but the proof is most abstract and non-constructive. Having gone through elaborate calculations and lists of conditions, the reader may appreciate the version that is quoted from a paper of (Hedberg [42]):

If $n \geqslant 2$ there exist real numbers $\lambda_1, ..., \lambda_n$ and elements $\varphi_1, ..., \varphi_{2n+1} \in C(I)$ with the following property: for each $f \in C(I^n)$ there exists $g \in C(R)$ such that

$$f(x_1, ..., x_n) = \sum_{q=1}^{2n+1} g[\lambda_1 \varphi_q(x_1) + ... + \lambda_n \varphi_q(x_n)].$$

The symbols $C(I)$, $C(I^n)$, and $C(R)$ denote, respectively, continuous functions defined on the unit interval, the $n$-dimensional unit cube, and the real line. The reader may consult Hedberg's paper for further detail.

Another novel approach to function representation that Doss extracted from the groundwork laid by Hedberg and Kahane used products of functions of one variable instead of sums. The statement of Theorem 1 in (Doss [29]) is centered on the representation

$$f(x_1, ..., x_n) = \sum_{q=1}^{2n+1} g \circ \prod_{p=1}^{n} \psi_{pq}(x_p) \tag{4.5}$$

and phrased in the language of Kahane's for 'quasi-all' functions $\psi_{pq}(x)$.

It can be brought into alignment with Theorem 5 by narrowing the scope by stating instead that 'there exist function $\psi_{pq}(x)$.

Doss observed that the product cannot be replaced by the sum in formula (4.5), but yet this version can be regarded a left-handed approach to representations with sums. Namely, applying the product decomposition in the case of functions of two variables gives

$$\psi_{1,q}(x_1)\psi_{2,q}(x_2) = \frac{1}{4}[\psi_{1,q}(x_1) + \psi_{2,q}(x_2)]^2 - \frac{1}{4}[\psi_{1,q}(x_1) - \psi_{2,q}(x_2)]^2$$

$$= h_1[\psi_{1,q}(x_1) + \psi_{2,q}(x_2)] + h_2[\psi_{1,q}(x_1) + \psi'_{2,q}(x_2)],$$

where $\psi'_{2,q}(x_2) = -\psi_{2,q}(x_2)$.

Formula (4.5) in the case $n = 2$ becomes a more complicated decomposition into sums of functions of one variable:

$$f(x_1, x_2) = \sum_{q=1}^{5} g \circ [\psi_{1,q}(x_1)\psi_{2,q}(x_2)]$$

$$= \sum_{q=1}^{5} g \circ \left\{ h_{1,q}[\psi_{1,q}(x_1) + \psi_{2,q}(x_2)] + h_{2,q}[\psi_{1,q}(x_1) + \psi'_{2,q}(x_2)] \right\}$$

A decomposition into sums can, of course, be extended to any $n$.

We have here another reminder of the economy and elegance of Kolmogorov's achievement.

Kolmogorov's function representations bring to mind several questions concerning their structure. These may be summarized as follows:

1. What functions $\psi_q(x)$ are admissible?

2. Are there admissible functions $\psi_q(x)$ for which the $g_q$ are unique?

3. Can the smoothness of $f$ and $g_q$ be correlated?

4. Is the number $2n + 1$ the least possible?

The first question received the most attention because of its central role in function representations (According to Kahane's existence Baire Category argument, almost all increasing functions would do, loosely speaking.) Functions whose sums separate the points of $E^n$ in the described manner in Theorem 2.1 were obtained as the limits of devil staircase functions. They are constructible to a degree, but inherently not computer friendly. They are one to one on large sets of $E^n$, and as such are singular: differentiable almost everywhere, with a zero derivative almost everywhere, and an infinite derivative on a dense set.

The second question appears to be more theoretical, and it received no attention at all; the answer is unknown.

The third question was answered by Vituškin in the sense of demonstrating that there is an inevitable descent in smoothness with a decease in the number of representation variables, and by further discoveries of Vituškin, Vituškin- Henkin, and Henkin.

The last question was established by direct geometric arguments by Doss for the case $n = 2$, and with dimension-theoretic arguments by Sternfeld in the general case.

There are two related questions that may override in importance for many readers the four that we had listed:

5. Can the representation of a specific target function be computed?

6. Can representations be developed into effective computational algorithms?

The answers to these questions are the subject of Chapter Four. We shall find that the answers are positive with caveats.

## 4.2 The anatomy of dimension reduction

The last chapter was concerned by and large with specific constructions and realizations of the ingredients that lay at the heart of Kolmogorov's function representations. We are turning the ring of our kaleidoscope half a turn to let in a different topological language along the Euclidean one. We already adjusted our kaleidoscope once before, when we discussed trees as indeterminate dimension-reducing agents in Chapter One.

Once again we anchor out thoughts to a commuting diagram:

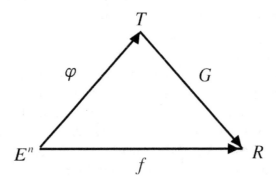

Figure 4.2: Commuting diagram

In keeping with the setting of function representations, this diagram describes schematically a real-valued function $f : E^n \to R$ with an alternative mapping in which an auxiliary function embeds the Euclidean cube $E^n$ in an indeterminate space $T$. A function $G$ computes $f$ on this space, expressed symbolically as $f = G(\varphi)$. This is by now a familiar method.

The dimension of the indeterminate space $T$ varies greatly in the pages of this monograph, from 1 to $2n+1$. An intermediate space of dimension 1 was the non- Euclidean universal tree in Chapter One; the reader might also think of T as standing for $n$ (or more) copies of $\Xi$. In this view, the dimension of $E^n$ was 'disentangled' into so many one-dimensional spaces (with a non- Euclidean topology), and the original dimension $n$ was restored with addition.

Kolmogorov achieved function representations with a scheme of remarkable economy and simplicity that leap-frogged over the dimensionality barrier using only functions of one variable and the binary operation '+' We have traveled the road that Kolmogorov took to achieve this result, and went through many computations and arguments, some heuristic, to demonstrate its validity.

Because of the equivalence of the functions $\psi_{pq}(x)$ and $\lambda_p \psi(x + qa)$ we consider here the representations (3.3). But regardless of the version, we have observed early on that the transformations that represented a given function with superpositions was a black box to us: We know the output of the black box, and we have a technique by which functions $g_1(y_1), g_2(y_2), g_3(y_3), \ldots$ compute an arbitrary function $f$ through a process of approximations. But we don't understand the mechanics of conversion from variables $x$ to variables $y$ other than symbolically. This proverbial black box is represented by the auxiliary function $\varphi$ in the diagram.

Thinking back about the elimination of coefficients in polynomial equations, from eight to three in the equation of degree 7 for example, this was done mechanically with arithmetic manipulations and radicals about which there was no mystery. Despite the tedious computations, that was a transparent procedure that could be followed step by step. The significant difference was that the elimination of coefficients involved discrete computations, without approximations and limiting processes. Transparency ceased when we stepped from algebraic into continuous operations.

The method of dimension reduction introduced by Kolmogorov was something else. Here we explicitly dealt with a geometric and analytic transformation of a Euclidean space to a lower dimensional one, each space containing uncountable many points.All of these considerations were irrelevant in the algebraic setting, and not even hinted at.

We did gain insight into dimension reducing function representations in Kolmogorov's 1956 paper. There the tree $\Xi$ was the indeterminate agency that reduced the $n$-dimensional cube to a one-dimensional configuration. An arbitrary continuous function $f(x_1, ..., x_n)$ could be represented in the form

$$f(x_1, ..., x_n) = \Sigma_{q=1}^{n} g_q(\xi_q)$$

with continuous functions $g_q$ of one variable, and functions $\xi_q$ that were also continuous functions of one variable, but albeit with values in a tree. The relation of a tree with level-sets made this scheme plausible, and we did not stop to get further insight into that process.

We interject at this point an important observation in understanding dimension-reducing function representations with superpositions is that every continuous mapping $f(x_1, \ldots, x_n)$ with $n \geq 2$ has non-trivial level sets:

$$L = \{(x_1, \ldots, x_n) : f(x_1, \ldots, x_n) = \text{const.}\}$$

whereas continuous mappings $t : E \to E^n$ have no such level sets (sets consisting of more than one point). These mappings are of two types: a tree $T$, and a space-filling curve $S$.

Kronrod had shown that level sets of a continuous function can be identified in a natural way with a tree; in this relationship their disjoint components are represented on different branches. With a space-filling curve, the points of level sets are dispersed and no such visual image is possible.

The one-one matching of the points $(x_1, \ldots, x_n) \in E^n$ for the points of $T$ or $S$ allows for organizing selected values of a target function $f$ onto stacks of layers of $T$ or $S$; with these the representation of $f$ is obtained continuously with addition.

The tree was Kolmogorov's first strategy for dimension-reducing superpositions and Arnold's strategy for refuting the conjecture in Hilbert's Problem 13; space-filling curve $s$ were the consequence of Kolmogorov's final strategy.

Kolmogorov's final representations formula included no external aids with a visual appeal such as a tree and level-sets. We are therefore faced with figuring out how sums of continuous functions of one variable $y_q = \Sigma_q \lambda_p \psi(x + qa)$ deal with the fact that for any two distinct points $(x_1, ..., x_n) \neq (x'_1, ..., x'_n)$ there are continuous functions at which $f(x_1, ..., x_n) \neq f(x'_1, ..., x'_n)$

Keep in mind that this must be true for *any* pair of n-tuples and *all* continuous function defined on $E^n$. The bare-bone presentation of Kolmogorov's 1957 paper is leaving the superpositions naked of imagery and make us wonder:

*What is the underlying mechanism of dimension-reduction in Kolmogorov's formula?*

*How are the sums of devil-staircase functions playing an equivalent dimension-reducing role as the tree ?*

We are led to assert that Kolmogorov's function representations exist only with superpositions that somehow order the points of the $n$-cube linearly the way the tree did. This is the puzzle that we are seeking to solve in the following pages, wondering also:

*Can computation-friendly function representations be achieved with functions other than nomographic functions?*

Before attempting to answer these questions, we need to look deeper into the covering property of the families of cubes $S_{q,k}(d_k)$ defined in Chapter Two. We remember that each function

$$\varphi_q(x_1, ..., x_n)$$

in the representation

$$f(x_1, ..., x_n) = \Sigma_{q=0}^{2n} g_q[\varphi_q(x_1, ..., x_n)]$$

maps continuously the unit cube $E^n$ onto an interval. These are special mappings in which, loosely speaking, almost all points of the cube are mapped one-one onto points of the range interval.[4] This fundamental feature of Kolmogorov's constructions merits further remarks.

Consider now one of the functions $\varphi = \varphi_q$. We already encountered the concept of *level set* when we discussed trees. It is the inverse image of a point $t$ in its range.

Generally, the equation $\varphi(x_1, ..., x_n) = t$ has infinitely many solutions for a given value of $t$, designated

$$\varphi^{-1}(t) = \{(x_1, ..., x_n) : \varphi(x_1, ..., x_n) = t\}$$

and this implies that there are points $(x_1, ..., x_n) \neq (x'_1, ..., x'_n)$ at which

$$\varphi(x_1, ..., x_n) = \varphi(x'_1, ..., x'_n).$$

This implies the obvious: that a single continuous function cannot separate all points of its domain. This, of course, implies that

$$g[\varphi(x_1, ..., x_n)] = g[\varphi(x'_1, ..., x'_n)]$$

for any function $g$ at the points $(x_1, ..., x_n) \neq (x'_1, ..., x'_n)$, but for any two such points there is a continuous function $f$ at which

$$f(x_1, ..., x_n) \neq f(x'_1, ..., x'_n).$$

---

[4]See Figure 2.

Therefore there must be functions $\varphi_q$ among the functions

$$\varphi_0, \varphi_1, ..., \varphi_{2n}$$

for which

$$\varphi(x_1, ..., x_n) \neq \varphi(x_1', ..., x_n')$$

at any specific pair of points. We know that the coverings used are such that there are $n+1$ values of $q$ for which this is true.

This leads to the inevitable conclusion that any iterative construction of superpositions that is based on coverings of the domain $E^n$ with pairwise disjoint cubes of diminishing diameters is subject to this point-separating property. This requirement shapes the behavior of the fixed inner functions in the sense of scrambling the linear order of the images of covering cubes. We noted how the function orders the images of the squares on the line real in the case of two variables.

Beware that $\varphi(\varphi^{-1})$ is the identity mapping of $E$ onto itself, presented symbolically as

$$\varphi[\varphi^{-1}(E)] = \varphi(E^n) = E,$$

but the mapping $\varphi^{-1}(\varphi)$ is not defined as a pointwise mapping of $E^n$ into $E^n$.

We are now in a position to look at the mechanism of dimension reduction underlying Kolmogorov's superpositions; we begin with Theorem 5 and the family of cubes families of cubes $\{S_{q,k}\}$. To analyze the mechanism whereby the inner functions $y_q = \Sigma_{q=0}^{2n} g_q[\varphi_q(x_1, ..., x_n)]$ enable the theorem, we assume that these families of cubes and functions are given, and for fixed $q$ and a given $k$ we examine the image intervals

$$\varphi_q(S_{q,k}) = \{t \in S_{q,k} : \varphi_q(x_1, ..., x_n) = t\}.$$

The sequence of pairwise disjoint cubes

$$S_{q,k_1}, S_{q,k_2}, S_{q,k_3}, ...$$

is mapped onto a sequence of pairwise disjoint intervals on the real line with a linear order dictated by the function

$$y_q = \varphi_q(x_1, ..., x_n).$$

This order can be discovered by computing the values of $\varphi_q(x_1, ..., x_n)$ at the lower left endpoint of the cubes. Such a situation of scrambled squares is shown in Figure 4.3.

Each cube can be mapped continuously onto its main diagonal in a variety of ways, and these diagonals can be connected in the order of their images to form a continuous curve $\Gamma_{q,k}$ with polygonal sections that lie in $E^n$ and do not intersect. Such a polygonal path is shown in Figure 4.4.

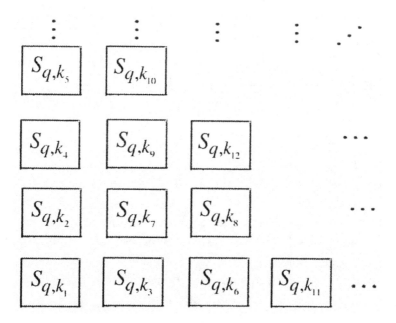

Figure 4.3: Example of ordered cubes $S_{q,k}$

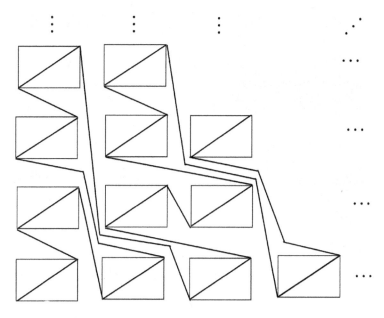

Figure 4.4: The polygonal curve $\Gamma_{q,k}$

With the linear ordering imposed on the squares $S_{q,k}$ by the functions $\varphi_q$, the curves connecting their diagonals converge to a continuous curve called *space-filling curve* $\Gamma_q$ as $k$ tend to infinity: $\lim_{k \to \infty} \Gamma_{q,k} = \Gamma_q$. We short-circuit the mathematical arguments that underlay the definition and construction of space filling curves, and simply state:

> $\Gamma$ is a space-filling curve if it is the graph of a continuous image of an injective mapping
>
> $$\Gamma : E \to E^n$$

(Sagan [82]).

It must be stressed that continuous mappings $\Gamma : E \to E^n$ and $\varphi : E^n \to E$ are <u>not</u> inverses of each other. From a computational perspective, the mathematical theory behind space-filling curve s as such plays no direct role in our actual computations, especially since in practice computations are limited to approximations with a few iterations. The existence of these curves as a limit is important, however, because they complicate the computation of superpositions through linear order of chains of cubes $S_{q,k_1}, S_{q,k_2}, S_{q,k_3}, \dots$ that they determine. These curves are therefore of central importance.

We turn now to the functions $y_q = \Sigma_{p=1}^{n} \lambda_p \psi(x_p + qa)$. There is a direct recursion routine that computes the function $\psi$, but not so these sums for fixed $q$: this computation involves combinatorial-type considerations and difficult algebraic number theory estimates due to the presence of the irrational numbers;[5] equivalent computational difficulties are also inherent in the sums $y_q = \Sigma_{p=1}^{n} \psi_{pq}(x_p)$.

The *a priori* linear ordering imposed on the cubes $S_{q,k}$ is transmitted to the computations of functions $g_q \circ \Sigma_{p=1}^{n} \lambda_p \psi(x_p + qa)$, and this is the computational Achilles heel of Kolmogorov's function representations.

The connection between this ordering and space-filling curve $s$ is analyzed through the approximating curves in (Braun [19]) and (Sprecher-Draghici [96]). Braun refers to this as the 'unfolding of dimensions'.

We illustrate this with the example of the function $f(x_1, x_2) = x_2$ shown in Figure 4.5. Its graph is a plane with the respective values $f\left(\frac{i_1}{10}, \frac{i_2}{10}\right) = \frac{i_2}{10}$ at the grid points. The difference in elevation between any two consecutive grid-points is 0 or $\frac{1}{10}$. We construct a function $g_q$ for a function representation

$$ f\left(\frac{i_1}{10}, \frac{i_2}{10}\right) \approx \sum_{g_q} \left[ g_q\left(\frac{i_1}{10}, \frac{i_2}{10}\right) \right]. $$

If $\varphi_q$ is as in Figure 4.5, than $g_q\left[\varphi_q\left(\frac{i_1}{10}, \frac{i_2}{10}\right)\right]$ is as pictured in Figure 4.6. This shows the impact of the space-filling curve $s$ on

---

[5]See, for example, Sprecher [85], page 348.

the computations. This demonstrates that the fluctuations of the functions $g_q$ are considerably worse than those of the target function $f$.

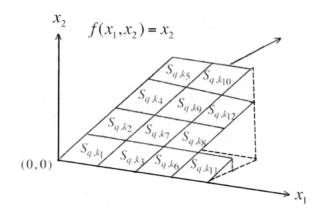

Figure 4.5: The function $f(x_1, x_2) = x_2$

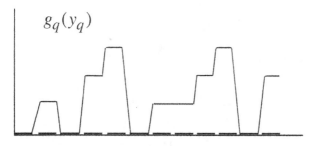

Figure 4.6: A function $g_q \left[ \varphi_q \left( \frac{i_1}{10}, \frac{i_2}{10} \right) \right]$

Two dimensions are easier to visualize than higher dimensions, and to make the subject more transparent we revert once again to Kolmogorov's function representations for functions of two variables. As noted earlier when doing this on similar occasions the arguments extend to functions of any number of variables in a straightforward manner.

A review of the literature since the 1970s shows numerous approaches and attempts of dealing with this pathology in implementing function representations. This refers, of course, to the scrambling of image intervals. A good example is the Boris Igelnik and Parikh use of cubic spline s [47]. The dimension-reducing algorithms of these functions carry the blame, and questions about their underlying structure from a mathematical perspective have been raised from the beginning.[6]

Let us look at grid-points in the linear order in which they are mapped by the functions $y = \psi_k(x_1) + \lambda \psi_k(x_2)$ onto the $y$-axis.

The graph connecting the grid-points

$$\left( \frac{i_1}{10}, \frac{i_2}{10} \right), \quad i_1, i_2 = 0, 1, ..., 10$$

for a constant $\lambda > 1$is shown in Figure 4.7. The situation changes markedly when we trace the grid-points,.

$$\left( \frac{i_1}{10}, \frac{i_2}{10} \right), \quad i_1, i_2 = 0, 1, ..., 10^2$$

Now the path zigzags jumps back and forth, as seen in Figure 4.7 and 4.8.

---

[6]See, for example Arnold [5], Lorentz [69], and Sprecher [90].

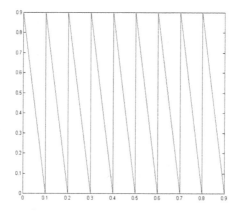

Figure 4.7: The path traced by $y = \psi_k\left(\frac{i_1}{10}\right) + \lambda\psi_k\left(\frac{i_2}{10}\right)$.

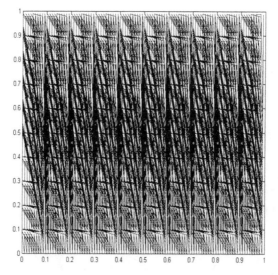

Figure 4.8: The path traced by $y = \psi_k\left(\frac{i_1}{10^2}\right) + \lambda\psi_k\left(\frac{i_2}{10^2}\right)$

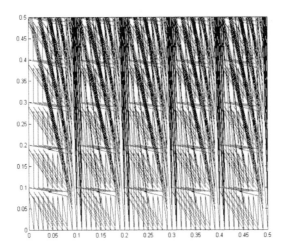

Figure 4.9: Detail of Figure 4.8

Now imagine the curves connecting the grid-points $\left(\frac{i_1}{10^k}, \frac{i_2}{10^k}\right)$, $i_1, i_2 = 0, 1, ..., 10^k$ as $k$ tends to infinity. Each of these limit curves can be represented in parametric form with a parameter $t$: $y = \zeta_k(t)$ with components

$$\begin{aligned} x_1 &= c_1(t) \\ x_2 &= c_2(t) \end{aligned} \quad t \in E. \tag{4.6}$$

Note that each of these curves can be regarded as a mapping from the interval $E$ into the square $E^2$, that is $\xi_k : E \rightarrow^{into} E^2$; like reversing the process of mapping the square into the line. A substitution of (4.6) into $y = \psi(x_1) + \lambda\psi(x_2)$ gives the function

$$y = \psi(c_1(t)) + \lambda\psi(c_2(t))$$

that is now a function of a single variable; it is evaluated along curves in the plane. With Figure 4.8 in mind, this says that space-filling curve $s$ have more than a ghostly presence in the computation of function representations.

A substitution of (4.6) into a function $f : E^2 \to R$ gives the representation

$$f(x_1, x_2) = \psi\left(c_1(t)\right) + \lambda\psi\left(c_2(t)\right)$$

showing that every continuous function of two variables is a continuous function of the parameter $t$ in the topology of space-filling curve $s$.

Generalizing this gives a rather novel example of reduction in the number of variables: For a given integer $n \geqslant 2$, let $\Xi$ be an arbitrary space-filling curve (such as the Hilbert space-filling curve ), generated with a continuous (surjective) mapping $\xi : E \to^{onto} E^n$ whose parametric representation $y = \zeta(t)$ has components

$$\begin{cases} x_1 = c_1(t) \\ x_2 = c_2(t) \\ \vdots \\ x_n = c_n(t) \end{cases} \quad t \in E$$

Then every continuous functions $f : E^n \to R$ has a representation

$$f(x_1, ..., x_n) = f(c_1(t), ..., c_n(t))$$

as a continuous function of the parameter $t$ in the topology of space-filling curve $s$. We shall discuss the consequences of these observations when we talk about computational algorithms.

These heuristic considerations point to the linearization that underlies Kolmogorov's strategies in the 1956 and 1957 papers. This is the process of organizing points $(x_1, x_2)$ along curves described with a single parameter. In general terms you might think of these curves as abstractions of the concept of a common curve that begin at a point on the boundary of the square $E^2$ (since we restrict this discussion to two variables) and end at another point on the boundary after passing through the points of $E^2$ without crossing itself and forming loops.

The discovery of space-filling curve s was the consequence of a theory that dealt with something else, and like tracing the origins of function representations, we must turn the calendar back over one hundred years.

In the second half of the 19th century, Georg Cantor was led to study infinite sets as a result of working on the problem of uniqueness of Fourier (trigonometric) series. This entailed comparing and classifying infinite sets in new ways and devising schemes for showing when elements of two infinite sets can be put in a one-one relationship. This led to the discovery that the unit interval $E$ can be put in a one-one relationship with the Cartesian product $n$-cube $E^n = E \times E \times ... \times E$ for any positive integer $n$.

This implied specifically the existence of a one-one (bijective) relationship $E \leftrightarrow E^n$ between the points of the unit interval $t \in E$ and the points on the $n$-cube $(x_1, ..., x_n) \in E^n$, a somewhat startling and counter-intuitive result, to say the least. That a bijective one-one mapping between the interval and a cube must necessarily be discontinuous was quickly proved by Eugen Netto, and this in turn raised the following question:

If instead of the one-one condition of a bijective mapping $E \leftrightarrow E^n$ we would consider the less stringent *injective* condition $E \rightarrow E^n$ that does not require a one-one relation, could such a relationship be continuous? That is: does there exist a continuous mapping $\Gamma : E \rightarrow E^n$ of the interval onto the cube?

This was answered in the affirmative by Giuseppe Peano, who discovered that there exist continuous curves passing through all points of a square or a subset thereof, thereby inducing a linear order on an infinite dense set of unordered points $(x_1, x_2)$. He was also the first to produce in 1890 such a curve. The graphs of such curves are called *space-filling curve s*. All this can be followed in (Sagan [82]).

Mathematicians since have followed his discovery with a detailed study of their properties as well as the construction of a rich variety of space-filling curve $s$; Hilbert himself produced one of the more intuitive such curves.

Reading these pages, I must admit that this capsule introduction to space- curves is quite obscure. The examples provide no guidance in trying to imagine the successive curves connecting points $\left( \frac{i_{1,k}}{10^k}, \frac{i_{2,k}}{10^k} \right)$.

To get into the spirit of things, therefore, we present as an example the Hilbert space-filling curve. It is one of the most intuitive and studied such curves and the process of constructing it geometrically easy to visualize, without the need to resort to intricate mathematical arguments. Unlike the examples above, the approximating curves to the Hilbert curve can be visualized and with a little effort traced on paper with a pencil. We shall indicate later how to use it to construct a computational algorithm as an alternative to the one based on Kolmogorov's function presentations.

Think of dividing the unit square $E^2$ successively into

$$4, 16, 64, ..., 2^{2k}, ...$$

sub-squares, an interval $E$ into as many equal segments [7]

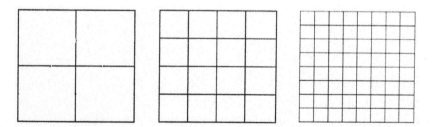

Figure 4.10: Partitions of the square.

---

[7]See Figure 4.10.

Hilbert did not attempt to prove the existence of space-filling curve $s \, \xi : E \xrightarrow{onto} E^2$. This was done by Giuseppe Peano ! He just wanted to construct one and he argued as follows: if the interval $E$ can be mapped continuously onto the square, then each line-segment lying in a sub-square can be mapped continuously onto the sub-square containing it.[8]

As $k$ increases, each sub-square is subdivided into four sub-squares, and this inclusion is preserved tends to infinity. The interval segments are always aligned to connect properly to the line segment of an adjacent square with a common boundary. Because every sub-square is mapped onto a line segment every point of the unit square will lie on the limiting curve. A complete description with technical detail and the inevitable mathematical constructions can be found in (Sagan [82]).

We now look at function representations in a new light. We know that the functions $y = \psi(x_1) + \lambda\psi(x_2)$ are the most basic nonlinear continuous functions that can be used to replace a pair of variables $(x_1, x_2)$ with new variables $y$, and satisfying the theorem: For the translations

$$y_q = \psi(x_1 + qa) + \lambda\psi(x_2 + qa) \qquad (4.7)$$

$q = 0, 1, 2, \ldots 5.$ there are continuous functions $g_q$ such that for any continuous function $f(x_1, x_2)$:

$$f(x_1, x_2) = g_0(y_0) + g_1(y_1) + \ldots + g_5(y_5).$$

According to the forgoing discussion, each of the functions (4.7) determined a space-filling curve so that these functions are evaluated along a one-dimensional curve and the dimension reduction of the function representations can be understood with these heuristic considerations.

---

[8]See Figure 4.11.

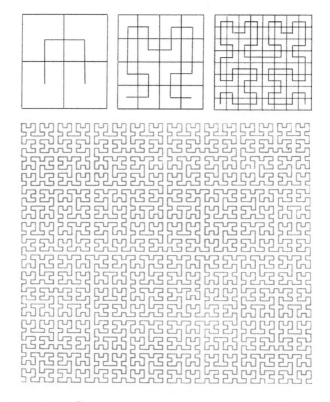

Figure 4.11: Successive approximations to the
Hilbert space-filling curve

## 4.3   Functional complexity

Up to this point the narrative intertwined research into analytic and
topological questions raised by Kolmogorov's function presentation
theorem. Dimension reduction falls mostly under the topology umbrella, as was amply evident as we followed the road to this remarkable and unexpected discovery. This remark reflects his 1956 paper
as well as Arnold's work that refuted Hilbert's conjecture and provided the link to the statement of function representation that was

shorn of these topological origins, but not entirely: The question of determining the minimal number of summands does retain topological considerations for values of $n$ beyond 2. Sternfeld regards Kolmogorov's theorem *"essentially a dimension-theoretic result."* [136].

The story line of this narrative paid little attention to the dimension-theoretic aspects raised by Kolmogorov's theorem and followed instead analytical questions. Prime among these was the structure and properties of the transfer function $\psi(x)$, and the resulting effects on the resulting function representations. Kolmogorov's theorem provided an answer to a specific analytic question in functional complexity. Here we explore briefly related questions within the orbit of this discovery.

Functions come in any number of variables, and in each case they are classified by their properties, such as continuity, differentiability, being analytic, trigonometric, and so on. Looking at the aggregate of all functions of a given number of variables, the question was if there were structural criteria that will distinguish between them. Hilbert launched this line of inquiry when he conjectured that some functions of three variables are distinguished from other functions of three variables. The measure that he used to define this distinction was the inability to represent them with superpositions of functions of two variables. He put no limit on the number of superposition that may be needed to show this, as long as there were only finitely many of them.

Hilbert was wrong in phrasing this concept in terms of continuous superpositions, but the question that he raised was based on a sound idea. Hilbert knew, for example, that there are analytic functions of three variables that are not representable with superpositions of analytic functions of only two variables. Vituškin demonstrated that the index combining differentiability with number of is a useful measure for classifying functions. This brings us to the general notion of finding measures for classifying functions of any number of variables that reflect complexity in some sense.

From this broad perspective, Kolmogorov's function representations are one measure among others. They resulted from the failed attempt to use the number of variables as a measure of complexity for the classification of continuous functions. Superpositions were the methodology according to which all continuous functions had measure 1. The concept of functional complexity reached beyond continuity to other measures of classification that can be considered. Its aim is the study of methods and numerical measures for the classification of functions.

We looked briefly at this concept in the Introduction, in connection with the interface of Problem 13 with Kolmogorov's function representations. Before moving the narrative to implementation and computing, we look at superpositions through this wider lens.

Consider for example the class of all functions having fourth order continuous partial derivatives. The equation

$$f(x_1, x_2, x_3) = g[\varphi_1 (x_1, x_2), \varphi_2 (x_2, x_3)]$$

would satisfy a fourth order equation (with 55 terms) (Buck [20]). This would determine the sub-class of functions $f$ representable in this form. The reader may recall Bochner's concept of generalized derivatives that was used earlier. With it, such representations could be studied even under weaker differentiability conditions. We mention this simply to point out questions and possible methodologies for answering them.

Without explicitly acknowledging it, we accepted the notion that combinations of functions of one variable are better in some sense than functions of several variables. This might be regarded as the overarching theme of this monograph.

In this same unarticulated sense, functions of fewer variables are more desirable than functions of more variables. In this language it can be said that Hilbert's conjecture was based on the idea that bad functions cannot generally be represented by good functions.

The trouble was that using continuity and the number of variables as an indicator were the wrong choices. This was a sound idea sidetracked perhaps by what might have been faulty phrasing

We were careful to verify function representations with uniformly convergent series of continuous functions. This guaranteed the preservation of continuity of the limit functions, but not necessarily preservation of the form. That is: if the uniform limit of the sequence of continuous functions

$$\lim_{k\to\infty} \sum_{q=1}^{N} g_{q,k}[\psi_{pq}(x_1, ..., x_n)]$$

exists, then we know that it is a continuous function, but we don't know if it is of the form

$$\sum_{q=1}^{N} g_q[\psi_{pq}(x_1, ..., x_n)].$$

The answer is unknown even in the case $N = 1$, as the following example shows:

The product function $x_1 \cdot x_2$ defined on the unit square $0 \leqslant x_1, x_2 \leqslant 1$ is the uniform limit:

$$x_1 \cdot x_2 = \lim_{k\to\infty} \exp\left[\ln\left(x_1 + \frac{1}{k}\right) + \ln\left(x_2 + \frac{1}{k}\right)\right]$$
$$= \lim_{k\to\infty} g_k[a_k(x_1) + b_k(x_2)].$$

The functions, $a_k$ and $b_k$ are continuous and strictly monotonic increasing; the product function is strictly increasing in each variable, except along the coordinate axes. Yet other than trivially, the limit is not of the form $g[a(x_1) + b(x_2)]$.

Vaĭnštaĭn and Kreines gave a partial answer to this convergence question in [104]. They found that the form of a sequence such as $g_k[a_k(x_1) + b_k(x_2)]$ is preserved when also the target function is strictly monotonic in each variable.[9] Arnold approached this problem from a broad point of view.

Remember Kolmogorov's first theorem in his 1956 paper that included the result that all continuous functions of two variables can be represented in the form

$$f(x_1, x_2) = g_1[\varphi_1(x_1, x_2)] + g_2[\varphi_2(x_1, x_2)] + g_3[\varphi_3(x_1, x_2)]. \quad (4.8)$$

All linear combinations of the functions in the right side are continuous, but the real-valued functions $\varphi_q(x_1, x_2)$ have values in the universal tree. Since these functions are fixed and independent of $f$, this says that the continuous functions $f(x_1, x_2)$ can be represented with continuous functions $g_q$ of one variable.

Arnold framed this result in a different language by saying that the combinations on the right side are *dense* in the space of continuous functions $f(x_1, x_2)$ defined on $E^2$. You might think of this as saying that all the functions $f(x_1, x_2)$ can be approximated with such linear combinations.

In contrast, he has shown that the simplest combinations

$$g[\psi_1(x_1) + \psi_2(x_2)]$$

are *nowhere dense* in the space of continuous functions. Loosely interpreting this statement, read it to mean that most continuous functions $f(x_1, x_2)$ cannot be approximated with continuous functions[10] $g[\psi_1(x_1) + \psi_2(x_2)]$.

It is interesting to note how much research was done on this simplest of superposition problems: [ Arnold [4], Ofman [75], Sprecher [90], Vaĭnštaĭn and Kreines [104] and others.

This interest is traceable to interest in nomography: the characterization of functions $f(x_1, x_2)$ that can be represented in the form $f(x_1, x_2) = g[\psi_1(x_1) + \psi_2(x_2)]$. The answer can be obtained

---

[9]See also Ofman [75].
[10]See V. P. Motornyi [72].

with a calculus routine when the functions have three continuous partial derivatives. Taking partial derivatives of both sides leads to the equation

$$\frac{\partial f}{\partial x}\frac{\partial f}{\partial y}\left[\frac{\partial f}{\partial x}\frac{\partial^3 f}{\partial x \partial y \partial y} - \frac{\partial f}{\partial y}\frac{\partial^3 f}{\partial x \partial x \partial y}\right]$$
$$+\frac{\partial^2 f}{\partial x \partial y}\left[\left(\frac{\partial f}{\partial y}\right)^2 \frac{\partial f}{\partial y}\frac{\partial^2 f}{\partial x^2} - \left(\frac{\partial f}{\partial x}\right)^2 \frac{\partial^2 f}{\partial y^2}\right] = 0$$

and this tells us which functions $f(x_1, x_2)$ are so representable.

These techniques and considerations carry us far from the geometric constructions that are the hallmark of Kolmogorov's function presentations.

The foregoing discourse and story line to this point focused on exact function presentations; that is, the focus of the research that we narrated was about equivalent representations of multi-variable functions with superpositions of fewer variables; one variable in the ultimate representations of Kolmogorov. But there is another sense in which the preference of reduction in the number of variables can be viewed, and this is process of approximation.

In practice, exact representations are seldom if ever achieved; they are carried out until a predetermined error is reached. The exact representation in the limit is important because it ensures that the error can be decreased at will, but this is not necessary if approximation is the goal.

Two complementary poles in functional complexity are Kolmogorov's function representation theorem, and Vituškin's theorem that ties dimension reduction with smoothness. In between these two poles are researches into alternative and more targeted representations such as we above. Straying away from exact representations is the question of *approximate representation*, in the sense of studying approximations with dimension reducing superpositions.

Kolmogorov's 1957 announcement overshadowed both his 1956 paper and Arnold's refutation of Problem 13. From the perspective of the goal achieved in the later paper, there was no incentive rather than a historic one to back to review discarded methodologies.

But the concluding considerations of this paper show that his thinking ranged beyond the common interpretations of that problem. This is evident from Kolmogorov's inclusion of the assertion that an arbitrary continuous function $f(x_1, ..., x_n)$ defined on $E^n$ can be approximated arbitrarily close by polynomial s of fewer variables of the form:

$$P(x_1, ..., x_n) = \sum_{r=1}^{2} a_r(x_n)b[c_r(x_n) + x_1, ..., c_r(x_n) + x_{n-1}]$$

We note that $b$ is a polynomial of $n - 1$ variables; the approach suggests the strategy of the first theorem of that paper from which his breakthrough followed.

To be more specific, Kolmogorov used in the proof of Theorem 1 $n + 1$ families of pairwise disjoint sets $G_{k,m}^1, G_{k,m}^2, G_{k,m}^3$ whose diameter tended to zero as $k$ tended to infinity, and certain non-negative functions $u_{k,m}^1, u_{k,m}^2, u_{k,m}^3$ that were constant on these sets. With coefficients $b_m^r$ depending on the sets $G_{k,m}^r$ and $f$, the function $f$ was approximated pointwise by functions

$$f(x_1, ..., x_n) = \sum_{r=1}^{n+1} \sum_{m=1}^{m_k} b_m^r u_{k,m}^r(x_1, ..., x_n).$$

Conceptually, this goes a step away from exact representations, and Kolmogorov acknowledged this in noting: *"This remark also illuminates from a rather new side the circle of problems relating to Hilbert's 13th problem."*

He can therefore be credited also with launching the concept of *approximate functional complexity* that was later developed by Buck [20, 21, 22].

This also connects with Bieberbach's early failed attempt to prove Hilbert's conjecture by asserting that not every polynomial of three variables of degree $k$ can be approximated with polynomial s of degree $k$ of two variables. [17,18]. *"it was not for nothing that L. Bieberbach called the 13$^{th}$ problem unfortunate"* commented Vituškin.

This concept concerns a variety of specific formats and as another means to define classes of functions. A case in point is the Weierstrass approximation theorem that tells us that uniform approximations (on $E^n$ in the context of this monograph) are always possible with polynomial s.[11]

In these considerations, the class of approximating functions (such as polynomials, Fourier series, cubic splines) is not related to the target functions being approximated, and preservation of format in the limit is not part of the language. An instance of this is approximations with nomographic functions that we already touched upon.

Another case in point is the innovative use of cubic spline s by Igelnik and Parikh [47] to which return in the next chapter. To take advantage of the reduction in the number of variables that is the hallmark of Kolmogorov's function representations without paying the price, they insert cubic spline s into the functions $\psi_k(x)$ and the construction of approximating functions $g_k(y_k)$. This makes Kolmogorov's formula into an effective computational device. It is a good approximators converging to target functions while foregoing the preservation of the format in the limit.

A classical problem that was touched on more than once is this: For what continuous functions $f(x_1, x_2)$ are there nomographic functions such that

$$f(x_1, x_2) = g[a(x_1) + b(x_2)]?$$

If a function $f$ cannot be represented in this form, how well can it

---

[11]See also [88].

be can it be approximated?

Interest in the general area of computing with dimension reducing algorithms was raised by the RAND Corporation and addressed in (Diliberto-Straus [27]). For a given error $\varepsilon > 0$ and continuous function $f(x_1, x_2)$ defined on $E^2$, they developed an algorithm for constructing continuous functions $g_1(x_1)$ and $g_2(x_2)$ such that[12]

$$\|f(x_1, x_2) - [g_1(x_1) + g_2(x_2)]\|$$
$$< \inf_{h_1, h_2} \|f(x_1, x_2) - [h_1(x_1) + h_2(x_2)]\| < \varepsilon.$$

[12]See also Buck [22] with other problems in approximate functional complexity.

# Chapter 5

# Hecht-Nielsen's neural network

## 5.1 Early beginnings

Reflecting on what we have covered so far, Kolmogorov's function representations were conceived in mathematical terms in response to a theoretical mathematical question. Computation was a frequently used term but mostly as a matter of speaking. Function representations were not intended as a computational tool, and we noted, in fact, inherent problems in their constructions that get in the way of efficient computing. Still, the combination of single variables functions and addition was attractive enough to try to overcome difficulties as they surfaced.

Recall the schematic representation of Robert Hecht-Nielsen's neural network in Figure 6. Interpreting it as a computer architecture shows the attraction of having a fixed intermediary layer that is stored irrespective of any application, and only five single variable functions that compute any target function in the case of two variables. Picking a large enough integer $n$, this set-up could be used to compute any continuous function of $2 \leqslant m \leqslant n$ variables

133

with minor adjustments for different values of $m$. With the universal function $\tilde{\psi}(x)$, the parameter $n$ can in principle be eliminated from this fixed layer.

The possibility of application to computations and application in areas outside of the mathematical sciences, including computer science, was lurking in the shadows from the time of publication of Kolmogorov's seminal paper; the absence of computational algorithms was a formidable barrier.

The mathematical interest in Kolmogorov's function representations was always within the confines of theory. Attempts to develop numerical constructions of superpositions before the late 1980s fell short of developing computational algorithms. In the view of this author, the state of mathematics and computer science up to that time was not ready for a crossover of this nonconstructive theory to application, and without that there was no point in going further.

Expectations of deep mathematical discoveries were disappointing. With informed insight acquired through his seminal work on entropy and generally approximation theory, Lorentz remarked as early as 1962 that Kolmogorov's theorem

> ... is very important in principle. Will it have useful applications? Theoretically, one could hope to derive by means of its results concerning functions of several variables from corresponding results about functions of one variable. For example, from the Weierstrass theorem about the approximation of functions of one variable by polynomial s at once follows the corresponding theorem for functions of s variables. This is not very astonishing, since most proofs of the theorem of Weierstrass generalize immediately to higher dimensions. One wonders whether Kolmogorov's theorem can be used to obtain positive results of greater depth. (Lorentz [66])

This was further echoed in the comment of Arnold who had, no doubt, the most profound knowledge of function representations:

> It is not clear to what extent the decompositions [ Kolmogorov's function representations ] can be further improved: for example, the question of uniqueness of the choice of the function $g$ has not been solved. Also there are no methods enabling one to represent a given smooth function as a superposition of functions that are also relatively smooth...

Arnold was referring to representations with a single function $g$

$$f(x_1, ..., x_n) = \sum_{q=0}^{2n} g \circ [\lambda_1 \psi_q(x_1) + ... + \lambda_n \psi_q(x_n)]$$

first published in (Lorentz [68]). If one could obtain such unique representations: representations in which for each $f$ there is a unique function $g$, this would have afforded a simple proof of the isomorphism of Banach spaces of continuous functions (Banach [6]).[1]

Arnold concluded as early as 1958 that

> we note that representations [Kolmogorov and Arnold formulas] are of purely theoretical interest, since they use essentially non-smooth functions...; therefore for practical purposes these representations are, it would seem, useless..." [3].

Lorentz came to regard them "of the nature of a pathological example, whose main purpose is to disprove hopes that are too optimistic" [69]; a theorem that stands alone like an edifice without

---

[1] A 1996-published proof by Miljutin [71] of the isomorphism of Banach spaces was actually achieved decades earlier. According to Vituškin, he was under a publication ban by Stalin, and his research was shelved in the archives of Moscow University. It came to light when its mathematicians went through the archives after Stalin's death.

offshoots. These were harsh judgments of the potential of Kolmogorov's discovery.

The comments reflected the prevailing view of most mathematicians who studied Kolmogorov's remarkable discovery closely. Research following its publication was aimed at structural questions concerning the anatomy of the functions of one variables that mapped the unit cube onto an $n$-dimensional surface in $2n+1$ dimensional space; the necessary number of summands; rate of convergence to the target function; and generalizations. The mathematical aspects of function representations and their depth had been pretty much plumbed during the three decades that followed its publication.

Other research that followed can be described as tying loose ends: adding transparency, tweaking the superpositions and investigating their mathematical characteristics, but it became clear that the structure determined by Kolmogorov's strategy did not allow significant changes or improvements.

Efforts of many provided various levels of insights, but with no numerical algorithms or applications, Kolmogorov's theorem remained being an existence theorem: Could it be the case from a mathematical perspective that Kolmogorov had said the last word on the subject, as the above comments implied?

In a constructive way he probably had as far as continuous functions were concerned, but the original deeper problem that triggered his discovery remained open: that of categorizing algebraic functions in terms of representations with analytic functions of fewer variables. Irving Kaplansky expressed the extreme view that superpositions of continuous functions had only second-order importance in the context of Problem 13.[2]

Yet, as if in defiance of these mathematical views, researchers began to test the viability of Kolmogorov's superpositions as a computational tool in pioneering efforts at locations two thousand miles

---

[2]See Appendix C.

apart: at the Los Alamos National Laboratory in New Mexico, and at the Courant Institute in New York. Bryan Travis experimented at Los Alamos between the late 1980s and early 1990s with a range of applications using the constructions

$$f(x_1, x_2) = \sum_{q=0}^{4} g_q \circ [\psi(x_1 + qa) + \lambda\psi(x_2 + qa)] \qquad (5.1)$$

in the Sprecher 1965 paper to explore a variety of computations in work that was never published; L. Frisch et al used the constructions

$$f(x_1, x_2) = \sum_{q=0}^{4} g \circ [\psi_q(x_1) + \lambda\psi_q(x_2)] \qquad (5.2)$$

in Lorentz's 1966 Approximation of Functions to compute global extrema.

We pointed out routinely that the discoveries centered on function representations were about mathematical existence, meaning that validity was established in principle, so to speak. Here were first efforts that derived computational algorithms from these existence theorems for actual applications. Both of these efforts demonstrated that enough information was embedded in either of these versions to show that function representations have the potential of being developed into an effective and sophisticated computational tool.

These researchers did not push the envelope beyond low-level iterations because their intent was to demonstrate the feasibility. They also showed the intrinsic effectiveness of the strategy that Kolmogorov devised, as judged by the rate of convergence in actual computing.

First to be noted is a striking coincidence in graphs of the function $y = \psi(x_1) + \lambda\psi(x_2)$ and $y = \psi_q(x_1) + \lambda\psi_q(x_2)$ shown in Figures 5.1 and 5.2 that Travis and Frisch et al had obtained:

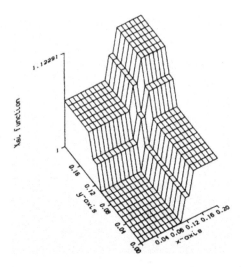

Figure 5.1: First iteration of the graph of $y = \psi(x_1) + \lambda\psi(x_2)$

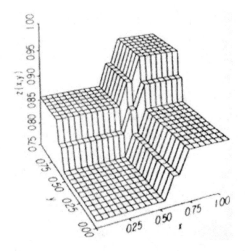

Figure 5.2: First iteration of the graph of $y = \psi_1(x_1) + \lambda\psi_2(x_2)$

These graphs are a general pictorial presentation of a very rough first approximation as the reader can surmise by looking back at the constructions of these functions. They were constructed by plotting the function values at grid points, and these graphs do not account for the gaps separating each of the depicted squares. More significant is the absence of elevation differential between squares in these graphs. If you think of the squares as seats in a stadium, then no two seats have the same height, for otherwise they would not be mapped into disjoint intervals. Still, these graphs do show the over all shape of these functions. (Compare these graphs with the colorful versions in (Dor Bar-Nathan [7] that came two decades later.)

The over-all goal of Travis ' explorations are summarized in his words:

> ... the point [of my computations] was to show that the K-S [ Kolmogorov - Sprecher ] algorithm can be applied not only to simple functions but just as easily to very complex functions, e.g., photographs of real-life scenes, or even pages of text. Computer speed and memory have increased enormously since these examples were generated over two decades ago. The K-S representation provides a base from which other applications, such as alternate methods of encryption, could be developed.

The first example computes a superposition representation of a function representing a tilted plane. Figure 5.3 shows contours of the target function and Figure 5.4 shows its superposition reconstruction This is a low-resolution reconstruction with a single iteration. Recalling Figure 3.6 and 3.7, you can see in Figure 5.5 close straddling of the first iteration of the original lines. A second iteration would have improved sharpness greatly, but the point here was simply to show the feasibility of reconstructing a function with representation Figure 5.5 graphs one of the outer functions in the superposition representation and Figure 5.6 shows a detail.

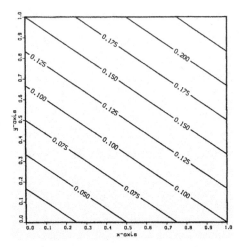

Figure 5.3: Tilted plane

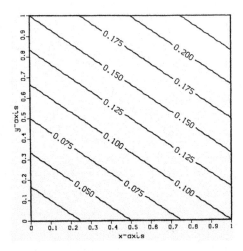

Figure 5.4: Reconstruction with one iteration

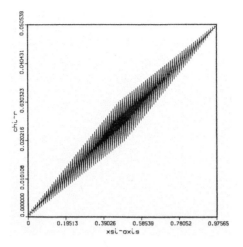

Figure 5.5: An outer function

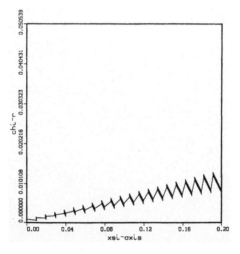

Figure 5.6: Detail

The next example is the one-iteration reconstruction of a bi-modal function in Figure 5.7. Figure 5.8 also displays the effectiveness of Kolmogorov's translation-scheme. The graph of one of the outer functions is depicted in Figure Figure 5.9.

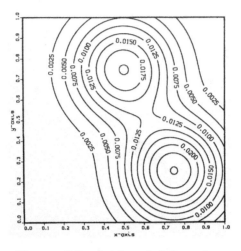

Figure 5.7: A bimodal function

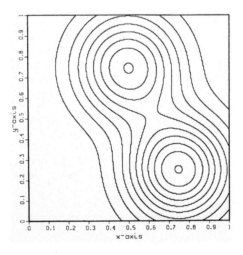

Figure 5.8: Reconstruction with one iteration

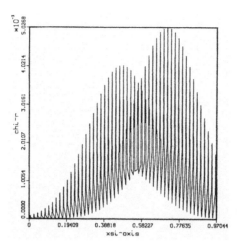

Figure 5.9: An outer function

The most significant of Travis ' applications was the reconstruction of an image of " Lena "[3] frequently used for comparison in tests of image compression. He applied Formula (5.1) to this image with the value $\lambda = 15$ to reconstruct the image in Figure 5.11. The result is impressive with this coarse gamma value and just one level of iteration. The graph of an outer function is shown in Figure 5.12.

Another level of iteration would greatly sharpen the image. Additional iterations were not carried out because Travis' point was to show that the K-S algorithm can be applied not only to simple functions but just as easily to very complex functions, e.g., photographs of real-life scenes, or even pages of text. The K-S representation provides a base from which other applications, such as alternate methods of encryption, could be developed. That so much information could be recovered even with a single iteration of function representations clearly established their potential.

---

[3]See Figure 5.1.

Figure 5.10: Lena

Figure 5.11: Reconstruction with one iteration.

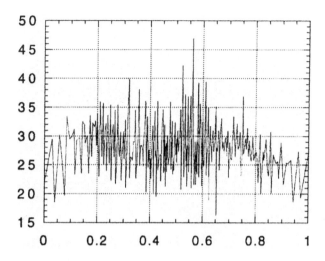

Figure 5.12: An outer function computing Lena

Of great interest about the underlying mechanism of Kolmogorov's function representations, and what enables the dramatic dimension reduction is the graph below. It might have pointed to the space-filling curve $s$ that enabled this dimension reduction, but another decade passed before this came into focus.

Figure 5.13 shows the second iteration of the unwinding path of the function $y = \psi(x_1) + \lambda\psi(x_2)$ for a small value of lambda It yielded a coarse set of squares, but it was easy to see the path that unwinds in the plane. To my knowledge this was the first attempt to visualize the order in which the images of grid-points of the unit square were arranged on an interval.

Arbitrary assumptions and experimental choices of values introduced computational errors as seen at the right side of the figure. A more refined 2002 approximation was shown in Figure 4.7.[4]

---

[4]See [96].

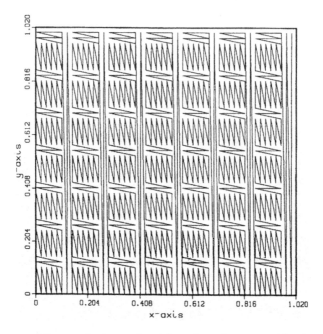

Figure 5.13: The ordering of grid points by $y = \psi(x_1) + \lambda\psi(x_2)$

Travis applied Kolmogorov's function representations to static problems. In contrast, Frisch et all applied them to a dynamic problem of finding global extrema. As the paper states:

> The principal point we make in this Letter is that, at least for functions of two variables for which we have already implemented the procedure, the approximation works surprisingly well with few iterations of [equation (5.5)]. The extension to more variables is straightforward. The criteria of goodness we are using are (1) the overall geometrical appearances of the reconstructed function, (2) the relative norm, $\theta_k$ of the error committed at the kth step, and (3) the local shifting of the extrema.[36]

As in the previous examples, the goal here was to establish the computational potential of superpositions rather than ultimate refinement. The target function is presented graphically in Figure 5.14. Its first approximation depicted in Figure 5.15 is indistinguishable in gross terms from the original. This is impressive for a single iteration, and in principle the authors algorithm is extendable to a wide range of functions of several variables.

The paper displayed also a fifth iteration that does not add new information.

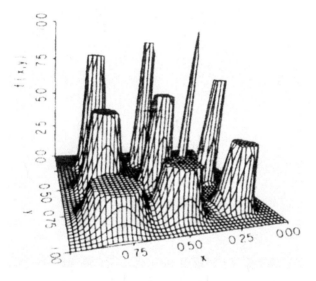

Figure 5.14: Source function

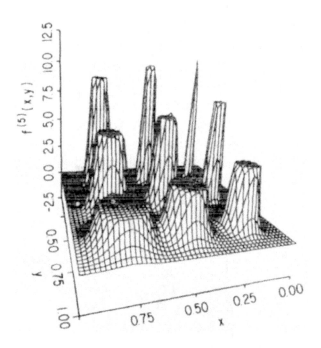

Figure 5.15: Reproduction with one iteration.

These may have been modest beginnings, but they were ground-breaking by demonstrating that function representations had the potential of an effective computational tool. Of particular importance was the rapid convergence that they exhibited, and the range of possible applications. In this regard, perhaps of particular notice was the reconstruction of Lena.

These successful efforts are particularly noteworthy in the context of the highly pessimistic assessment of mathematicians. We noted that Arnold, in particular, saw no promise as late as 2004 – twenty five years after Travis showed that image compression was accessible to superposition computations, and Frisch el all published their findings.

# 5.2 Hecht-Nielsen's neural network

The ground-breaking applications of Kolmogorov's function representations, one unpublished, broke the ice but still representations remained to be regarded as mathematical constructs. A change in paradigm was needed to move them from their mathematical formulation to computational algorithms.

Enter computer scientist Robert Hecht-Nielsen. Acknowledging these reservations, he observed:

> Although Kolmogorov's theorem was both powerful and shocking (many mathematicians do not believe it can be true when they first see it), it has not been found to be of much utility in terms of its use in proving other important theorems. In mathematical terms, no one has found a significant use for it.

But his next sentence was a rite of passage from the mathematical cradle to applications:

> 'The point of this paper is that this is not the case in neurocomputing!

In view of the pessimistic assessments and reservations, including his own, this represented a remarkable optimism, in particular in view of Arnold's verdict. Beyond this it showed a profound insight that for the first time framed this mathematical existence proposition in a broader setting of another existence that opened the door to actual implementation. This was a watershed statement that gave superpositions a new interpretation.

The backdrop of Hecht-Nielsen's statement was the pivotal formula in the preface that is repeated here in the equivalent form:

$$\begin{cases} y_q = \sum_{p=1}^{n} [\lambda^p \psi(x_p + q\varepsilon) + q] \\ f(x_1, \dots, x_n) = \sum_{q=0}^{2n} g_q(y_q) \end{cases} \tag{5.3}$$

As we commented earlier, this displays more clearly the two distinct stages of the computation: The fixed functions $y_q$ are computed for a given natural number $n$, and then a target function $f$ is computed.[5] We recall, however, that this is an iterative process in which approximating sequences to $f$ are constructed on approximating sequences to the functions $\psi$ and hence $y_q$.

It was this version of Kolmogorov's function representation formula that Hecht-Nielsen interpreted as a feed-forward neural network in a brief note titled: *Kolmogorov's Mapping Neural Network Existence Theorem*. It was presented at the First IEEE International Conference on Neural Networks held in San Diego, California in June, 1987. With a slight change in format and alternative notation for easier comparison with the notation of the preceding sections, Hecht-Nielsen's verbatim statement stated the following:

Hecht-Nielsen's Kolmogorov's Mapping Neural Network Existence Theorem.

**Theorem 8.** *Given a real-valued continuous functions*

$$f : E^n \to R, f(x_1, ..., x_n) = z,$$

*then $f$ can be implemented exactly by a three-layer feed-forward neural network having $n$ fanout processing elements in the first (input) layer, $(2n + 1)$ processing elements in middle layer, and $(2n + 1)$ processing elements in the output layer.*

*The processing elements in the input layer are fanout units that simply distribute the input components $(x_1, ..., x_n)$ to the processing elements of the second layer.*

*The processing elements of the second layer implement the following transfer function:*

$$y_q = \sum_{p=1}^{n} [\lambda^p \psi(x_p + q\varepsilon) + q]$$

---

[5]See Figure 6.

*where the real constant $\lambda$ and the continuous real monotonic increasing function $\psi$ are independent of $f$ (although they do depend on $n$) and the constant is a rational number $0 < \varepsilon \leqslant \delta$, where $\delta$ is an arbitrarily chosen positive constant.*

*The top $m$ $[m = 2n + 1]$ layer processing elements have the following transfer functions:*

$$\sum_{q=1}^{2n+1} g_q(y_q)$$

*where the functions $g_q$, $q = 1, 2, ..., m$ are real and continuous and depend on $f$ and $\varepsilon$.*

This insightful formulation was equivalent to Kolmogorov's Theorem in a setting that had to be reconciled with the elements of neural networks. It made a bold connection to computer architecture, but in concept only and not as an immediate practical neural network poised for implementation. As Hecht-Nielsen summarized the situation:

> No specific example of a function and a constant are known. No example of a $g$ function is known. The proof of the theorem is not constructive, so it does not tell us how to determine these quantities. It is strictly an existence theorem. It tells us that such a three-layer mapping network must exist, but it does not tell us how to find it. Unfortunately, there does not appear to be much hope that a method for finding the Kolmogorov network will be developed soon. Thus, the value of this result is its intellectual assurance that continuous vector mappings of a vector mapping of a vector on the unit cube (actually the theorem can be extended to apply to any compact i.e., closed and bounded, set) can be implemented exactly with a three-layer neural network.

He stated further:

> Kolmogorov's Mapping Neural Network Existence The-
> orem is a statement that our quest for approximations
> of functions by neural networks is, at least in theory,
> sound. However, the direct usefulness of this result is
> doubtful, because no constructive method for develop-
> ing the functions is known. [40]

These statements are quoted at length to underscore the shrewd insight that Hecht-Nielsen has shown in the face of prevailing negative
assessments of implementing Kolmogorov's superpositions. The repeated use of the term *existence* underscores the strong reservations
about actual implementation.

Starting with the format that suggested a computer architecture, Hecht-Nielsen analyzed the elements of the formula and concluded that the transition from a mathematical construct to a computational tool had a sound theoretical basis. He recognized that
this was only in principle, since this model had yet to be put to
the test.

Indeed, theoretical and practical criticism of Hecht-Nielsen's
vision was quick to follow on the heels of its dissemination. The
topology of Hecht-Nielsen's feed-forward neural network that was
suggested by the modified formula not withstanding, the functions
$\psi$ and $g$ were not of the type used in neurocomputing; in addition, $g$ was dependent on the target function $f$ and thus was not
representable in suitable parametric form.

These factors raised questions about the feasibility of actual
implementation of the formula as a neural network: Was Hecht-
Nielsen's network interpretation valid, and was it really relevant
to neurocomputing? After all, Kolmogorov's function representations were an existence theorem, and so was its Hecht-Nielsen's
interpretation as a neural network. Tellingly, either as a mathematical construct or as a neural network, no part of the formula
was computable.

It is interesting to note that these pros and cons were argued in disregard of the ad hoc success of using two different versions of Kolmogorov's formula as a computational tool. This is not surprising in the context of different research aims and methodologies, and the diversity of the research community. Anyway, Travis did not disseminate his findings.

The feasibility of applicability of function representations to computations on the one hand, and implementation as a neural network on the other, exposed the wide gap between Kolmogorov's and Hecht-Nielsen's two existence theorems and the possibility of transforming dimension-reducing superpositions into workable algorithms compatible with neural networks. Hecht-Nielsen himself articulated these reservations and doubts, and we note that his theorem was intended to aid in understanding neural networks.

This gap was narrowed over the next four years, first surprisingly by the sharpest critics of Hecht-Nielsen's concept: Federico Girosi and Tomaso Poggio. After rejecting the exact representation that was the hallmark of Kolmogorov's theorem and Hecht-Nielsen's interpretation, however, they suggested approximation as a possible alternative application of superposition to computing.

A good bridge between the mathematical construct of function representations and neural networks at that point in time can be found in papers of Kůrková [59-61], H. L. Frisch et al who was cited above, and Nakamura et al [73].

*Kolmogorov's Theorem: An Exact Representation is Hopeless* was the response by Girosi and Poggio in *Neural Computation* [37]. Their argument, under the title: *Representation Properties of Networks: Kolmogorov's Theorem is irrelevant* did have credibility from the perspective of actual implementation.

They listed learning—the basic feature of neural networks, and approximation as two specific reasons: The first was that implementation required a certain degree of smoothness of the functions $\psi$ and $g_q$ to stabilize computations and enable filtering out noise; the known inner function $\psi$ was pathological. They cite the inevitable descent in dimension-reducing smoothness of superpositions proved by Vituškin, and related limitations by Henkin as described below.

The second reason was the need for parameterization that could not be met because of the dependence of the functions $g_q$ on the target function $f$. They also pointed to parallel complexities between the functions $g_q$ and the target function that they compute:

> Peripherally connected to these technical and theoretical arguments concerning the application of function representations to neural networks, were general mathematical considerations concerning their usefulness in furthering the study of continuous and smoother functions.

In this Girosi and Poggio were not alone. They provided a good connection to theoretical considerations made twenty-five years earlier in a strictly mathematical context, such as Arnold's and Lorentz's questioning if Kolmogorov's theorem might have "useful applications" or "obtain positive results of greater depth".

This was further enhanced in the second part of the Girosi - Poggio paper that contains an insightful discussion of Hilbert's problem 13. Despite their conclusion that *"An Exact Representation is Hopeless...,"* they did leave open the possibility of approximate representations.

This paper was important in that it focused on the difficult conciliation between theory and practice, and launching a fruitful discussion. This has been a thorny problem all along, address obliquely rather than directly. Mathematicians by and large focused on theory, and once the existence of Kolmogorov's function representations was established and failed to yield deeper mathematical insights, lost interest.

A 1972 survey article by the author underscores this point [90]. Listing the state of development of function representations up to that time, it closed with questions about alternative conditions to guarantee their existence. The possibility of application to computation was not even eluded to indirectly. It was simply not part of the mathematical interest. It took the early beginnings described above and Hecht-Nielsen's interpretation to shift the focus away from existence to numerical realization of superpositions and computations.

To get us all on the same page it is appropriate to digress for a brief summary of the nature of computing being discussed here, and its nomenclature:

The Poggio- Girosi criticism appeared in print within two years of Hecht-Nielsen's announcement. Two years after that, Kůrková responded to it with a paper titled *Kolmogorov's Theorem is Relevant* [60]. She restated in broader detail their concerns, observing that the formula is reminiscent of a perceptron-type network despite the noted misfit between the component functions and neural network demands.

She stated her remarks in the context of a multilayer network along the lines of Figure 6, where weighted inputs are summed up in the middle layer, and weighted inputs are summed up in the output layer, and then argues that the functions in the formula are suited for staircase like functions of sigmoidal type. Following a discussion of the mechanics of Kolmogorov's strategy, Kůrková offered the following proposition in that paper:

The building blocks are functions $\sigma(t)$ of sigmoidal type. At the risk of being redundant I mention for the benefit of readers not familiar with neural networks that a sigmoid function is a bounded $S$ shaped function defined at all points of the real line R with a variety of formulas. An example is the logistic function

$$\sigma(t) = \frac{1}{1 + e^{-1}}.$$

It is easily verified that it tends asymptotically to 1 as $t$ tends to infinity, and to zero when $t$ tends to minus infinity:[6]

$$\sigma(t) : R \to E, \qquad \lim_{r \to -\infty} \sigma(t) = 0$$

and

$$\lim_{r \to \infty} \sigma(t) = 1$$

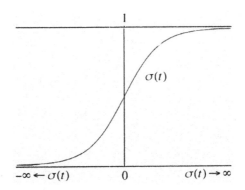

Figure 5.16: A sigmoidal function

These functions can be used to build staircase-like functions $\psi_{pq}$ and $g_q$ such that the functions

$$\tilde{f}(x_1, ..., x_n) = \sum_{q=0}^{N} g_q \circ \sum_{p=1}^{n} \psi_{pq}(x_q)$$

approximate a given continuous function $f(x_1, ..., x_n)$.

Stated more precisely, this can be done once a function $f(x_1, ..., x_n)$ and a small error $\varepsilon > 0$ are given there is a positive integer $N$ and staircase-like functions of sigmoidal type $\sigma$ and functions $\psi_{pq}$ and $g_q$ such that[7]

$$\max_{E^n} \left| f(x_1, ..., x_n) - \tilde{f}(x_1, ..., x_n) \right| < \varepsilon. \tag{5.4}$$

---

[6]See Figure 5.16.
[7]See H. Chen-T. Chen- R. Liu [23].

Under similar conditions, Kůrková also obtained a representation

$$f(x_1, ..., x_n) = \sum_{q=0}^{N} g \circ \sum_{p=1}^{n} w_{pq} \psi_q(x_p),$$

with constants $w_{pq}$ under weak conditions. The functions $\psi_q$ are uniform limits of sequences of staircase-like sigmoidal-type functions whose continuity and monotonicity is not required. These theorems suggest the feasibility of Hecht-Nielsen's network idea, and they narrowed the gap between his existence network and computer implementation. Error estimates[8] for any choice $\varepsilon > 0$ could be achieved but at the cost of increasing the number of neurons as $\varepsilon \to 0$.

Returning to the formula:

$$f(x_1, ..., x_n) = \sum_{q=0}^{2n} g_q \circ \sum_{p=1}^{n} \lambda^{p-1} \psi_q(x_p + q\varepsilon), \qquad (5.5)$$

the gap eluded to was narrowed further in a paper *Computational Aspects of Kolmogorov's Superposition Theorem* [70] that appeared a year later following Kůrková's lines of reasoning.

A continuous function $0 \leqslant \sigma(t) \leqslant 1$ such that[9]

$$\sigma(t) = \begin{cases} 0 \; when \; t \leqslant 0 \\ \\ 1 \; when \; t \geqslant 1 \end{cases}$$

and a sequence $\{\lambda_k\}$ of integrally independent natural numbers are given. Then there exist $\lambda_k$-dependent constants $a_{k,i}$ for which the sequence

$$\psi_k(x) = \sum_{i=1}^{m_k} a_{k,i} \sigma(k!x - i) \qquad (5.6)$$

---

[8]See also Hornik -Stincome [46] and Blum-Li [15].
[9]See Figure 5.17.

converges to a continuous function $\psi(x)$ on the interval $0 \leqslant x \leqslant 1 + \frac{1}{5!}$.

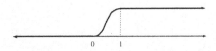

Figure 5.17: The function $\sigma(t)$

Observe that equation (5.6) represents for each value of $k$ a sum of translations of the function $\sigma(k!x - i)$, where the coefficients $a_{k,i}$ represent height. The graphs of the functions $\psi_1(x)$ and $\psi_2(x)$ are given in Figures 5.18 and 5.19, respectively.

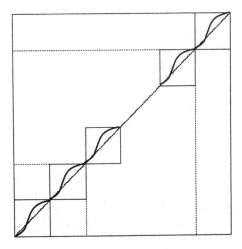

Figure 5.18: The graph of $\psi_1(x)$.

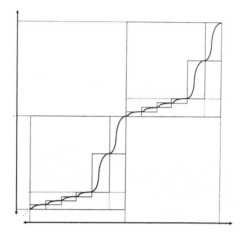

Figure 5.19: The graph of $\psi_2(x)$.

This is the third method by which the function $\psi(x)$ was being explicitly constructed in this monograph, and it might have been more properly included earlier. It is included here because with it the representation formula (5.6) can be proved with a novel technique. This is done in the corollary below:

**Corollary 1.** *For every continuous function $f : E^n \to R$ for $n \geqslant 2$, the continuous functions $g_q(y_q)$ in formula (5.6) with constant $\varepsilon = \sum\limits_{r=n+2}^{\infty} \frac{1}{r!}$ can be derived as follows: There exist constants $c_s$ and $\xi_{s,j_{q_1},\ldots,j_{q_n}}$ depending on the constants $\lambda_1, \ldots, \lambda_n$ and $\psi$, and constants $b_{s,j_{q_1},\ldots,j_{q_n}}$ depending on $f$, such that the series of functions*

$$g_{q,r}(y_q) = \sum_{s=1}^{r} \sum_{s,j_{q_1},\ldots,j_{q_n}} b_{s,j_{q_1},\ldots,j_{q_n}} [c_s(y_q - y_{s,j_{q_1},\ldots,j_{q_n}}) + 1]$$

*converges uniformly to $g_q(y_q)$ for $q = 1, 2, \ldots, 2n + 1$.*

These constructions appear to be the more intricate and daunting than earlier ones, but there is a significant payoff: sigmoidal functions enter the computations. Specifically, let $\hat{\sigma} : R \to E$ be a sigmoidal function, and $S(\hat{\sigma})$ the set of functions $\sum_{s=1}^{r} a_s \hat{\sigma}(b_s x + c_s)$ with constants $a_s$, $b_s$, and $c_s$, then the functions $\psi_k(x)$ and $g_q(y)$ can be approximated with functions $\hat{\psi}_k(x)$ and $\hat{g}_q(y)$ in the set $S(\hat{\sigma})$ with a predetermined error. Kůrková obtained the resulting approximation formula above within the constraints imposed by neural networks application (Kůrková [61]).

The method for achieving the above alternative proofs is called *contraction mapping s*. For our limited purposes here, we can think of this class of mappings in terms of the *Lipschitz continuity* example that we already encountered: $|f(x) - f(x')| \leqslant A |x - x'|$ for some constant $A$ and any two points in its domain. For this to be a contraction mapping, we put $A = 1$. Thus: if $f : E \to R$, then the height of the rectangle $|f(x) - f(x')|$ never exceeds the width $|x - x'|$ of the base for any two points $x$ and $x'$ in $E$. This is a technique for scaling copies of one of the squares in Figure 1 and placing it into each of the boxes in that figure, and iterating this process.

The actual set-up of a contraction mapping is by its nature elaborate, and we do not pursue it. The reader can find a detailed construction in (Katsuura-Sprecher [51]).

Another paper making this point more analytically but without a specific application is the Nakamura et al *Guaranteed Intervals for Kolmogorov's Theorem (and their possible relation to Neural Networks)* [73]. Recall that the function $\psi$ in formula is the uniform limit of sequences $\psi_k$ whose prime property is to map pairwise disjoint intervals of length $\delta_k = \frac{\gamma-2}{\gamma-1}\gamma^{-k}$ into pairwise disjoint image intervals of length

$$\varepsilon_k = (\gamma - 2) \sum_{r=1,2,3,\dots} \gamma^{-\beta_k + r}$$

where $\gamma \geqslant 2n + 1$ is fixed in advance for a given $n$, and $\beta_k \geqslant (n+1)^{k-1}$.

These definitions are what the authors use. To fixe ideas, this paper can be thought of in terms of a recurring classical statement known to any student of calculus: Given an error $\varepsilon > 0$ there exists a number $\delta > 0$ such that if $|x - x'| < \delta$ then $|\psi(x) - \psi(x')| < \varepsilon$. This is roughly how continuity of $\psi$ at a point $x$ would be tested. The trouble with this test is that we know that such a number $\delta$ exists, we are generally unable to find it. This is a serious problem if we wish to carry out an approximate computation to within a given error $\varepsilon > 0$.

Using the positive results in the papers of Frisch et al [46] and Nees e [74], the authors observe that even though they establish rapid convergence of approximations, accuracy cannot be fixed in advance. This paper attempts to construct approximation algorithms that would compute the intervals and their unique images that is possible in principle, but not in practice. The authors present a sequence of choices, and if these could be streamlined, then the functions $g_q(y_q)$ that compute a target function $f$ might be suitably approximated.

This intent, to provide an algorithm with a guaranteed accuracy, is an important step to make the technique of superpositions a practical computational algorithm. Yes there is an irony here, because, as the author's concede, this guaranteed accuracy was at that stage an existence rather than an implementable algorithm.

## 5.3   Computational algorithms

With the exception of the ad hoc applications of Bryan Travis and Frisch et al to computing, the story that we followed was rooted in theory. Hecht-Nielsen's shift of paradigm and the discussions that followed intensified interest in implementing function representations. This entailed the development of numerical algorithms for the construction of the function $\psi(x)$, and a concurrent algorithm for computing the sums

$$y_q = \lambda_1 \psi_1(x_1 + qa) + ... + \lambda_n \psi_n(x_n + qa)$$

and the functions $g_q(y_q)$ that compute target functions.

Further problems had to be overcome before superposition formulas could be adapted to neural networks.

Other than the difference in the number of terms and elements, there is no tangible difference between the algorithm for computing functions of $n$ variables or functions of two variables. Readers new to the subject who wish to use the algorithm for a specific target function may benefit from the case $n = 2$, and this is the setting of the first algorithm to be described.

The algorithm can be thought of as a three-layer network with an input, middle, and output layer fashioned on Robert Hecht-Nielsen's neural network. [126][10] It is in the nature of a model that needs adjustment to conform to neural network requirements, as we have observed earlier. Mario Köppen generalized this algorithm for general computing [57].

The computational algorithm is presented in the decimal base:

$$d_k = \sum_{r=1}^{k} \frac{i_r}{10^r}$$

$i_1 = 0, 1, ..., 10$, and $i_r = 0, 1, ..., 8$ otherwise.

The specific function representations are:

---

[10]See Figure 6.

$$\begin{cases} f(x_1, x_2) = \sum_{q=0}^{4} g_q \circ h_q(x_1, x_2) \\ h_q(x_1, x_2) = \psi(x_1 + qa) + \lambda\psi(x_2 + qa), \ q = 1, 2, 3, 4 \\ \lambda = \sum_{r=1}^{\infty} \frac{1}{10^{\beta(r)}} \\ \beta(r) = 2^r - 1 = 1 + 2 + \dots + 2^{r-1} \\ a = \frac{1}{9 \cdot 10} = \sum_{r=2}^{\infty} \frac{1}{10^r} \end{cases} \quad (5.7)$$

This notation has already been used in this monograph, and should be familiar to the reader.

## 5.4   The first algorithm

The starting point is a function $\sigma(x)$ defined on the entire real line as depicted in Figure 5.20.

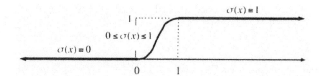

Figure 5.20: The function $\sigma(x)$.

This prototype is used to define functions[11] $\theta_{q,k}(y)$:

$$\theta_{q,k}(y) = \sigma\left[10^{\beta(k+1)}[y - h_q(d_{1,k}^q, d_{2,k}^q) + 1]\right]$$

$$-\sigma\left[10^{\beta(k+1)}[y - h_q(d_{1,k}^q, d_{2,k}^q) - 8b_k]\right] \quad (5.8)$$

---

[11]See Figure 5.21.

where

$$b_k = (1 + \lambda) \sum_{r=k+1}^{\infty} \frac{1}{10^{\beta(r)}}$$

It is important to bear always in mind that these functions are defined for each pair of grid-points $(d_{1,k}^q, d_{2,k}^q)$

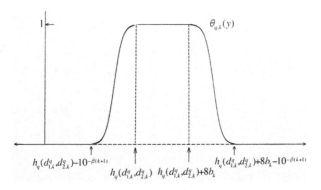

Figure 5.21: The function $\theta_{q,k}(y)$

These are the computational components. Note that the function $\theta_{q,k}(y) \cdot f(d_{1,k}^q, d_{2,k}^q)$ has the shape as in this figure, except that its height is $f(d_{1,k}^q, d_{2,k}^q)$; if this value is negative, then the graph would be inverted.[12]

Defining $f_0 \equiv f$, assume that the function $f_{r-1}(x_1, x_2)$ is known. Select an error $\varepsilon > 0$ and determine a large enough integer $k_r$ for which

$$|f_{r-1}(x_1, x_2) - f_{r-1}(x_1', x_2')| \leqslant \varepsilon$$

when

$$|x_p - x_p'| \leqslant \frac{1}{10^{k_r}}$$

for $p = 1, 2$.

---

[12]See Figure 3.8.

This is reminiscent of an often repeated calculus argument: For example, the function $g(x)$ is continuous at the point $x_0$ if for a given an error $\varepsilon > 0$, there exists a number $\delta > 0$ such that $|g(x_0) - g(x)| < \varepsilon$ when $|x_0 - x| < \delta$. As a rule, you were not asked to find that $\delta$, and it is not always possible to find it. That raises the question if it is possible in practice to find the integer $k_r$ that will guarantee the desired accuracy, for otherwise the algorithm may not be of practical relevance.

We have encountered this question in connection with Frisch et all in Section 5.1. The results of Nakamura et all in [73] on this subject apply also here.[13]

Input Layer

$$\boxed{d^q_{1,k_r}, d^q_{2,k_r}}$$

You input in this layer the translated decimal grid-points $d^q_{1,k_r} = d_{1,k_r} + qa$ and $d^q_{2,k_r} = d_{2,k_r} + qa$ using the number $k_r$ just determined.

Middle (hidden) layer

$$
\boxed{
\begin{array}{c}
\theta_{q,k}(y) \\[4pt]
\uparrow \\[4pt]
h_q(d^q_{1,k_r}, d^q_{2,k_r}) = \psi(d^q_{1,k_r}) + \lambda\psi(d^q_{2,k_r}) \\[4pt]
\uparrow \\[4pt]
\psi(d^q_{1,k_r}), \psi(d^q_{2,k_r})
\end{array}
}
$$

Perform these calculations for the values $q = 0, 1, 2, 3, 4$. First the function is computing on the $x_1$ and $x_2$ axes respectively at the grid-points. This is followed by computing the sums that are

---

[13]See also Trofimov-Kariton [102].

formed with the multiplier $\lambda$. Finally the function $\theta_{q,k}(y)$ is computed at the valued $h_q(d^q_{1,k_r}, d^q_{2,k_r})$ computed in the previous step. These calculations do not involve any function of two variables to be computed.

### Output Layer

$$f_r(x_1, x_2) - f(x_1, x_2) = \sum_{q=0}^{4} g_{q,r} \circ h_q(x_1, x_2)$$

$$\uparrow$$

$$g_{q,r} \circ h_q(x_1, x_2) = \sum_{d^q_{p,k_r}} \tfrac{1}{3} f_r(x_1, x_2) \cdot \theta_{q,k_r}[h_q(x_1, x_2)]$$

$$\uparrow$$

$$g_{q,r} = \sum_{d^q_{p,k_r}} \tfrac{1}{3} f_{r-1}(x_1, x_2) \cdot \theta_{q,k_r}(y)$$

These are the computations that evaluate the $r^{\text{th}}$ approximation to the function $f(x_1, x_2)$ of two variables. This completes the $r^{\text{th}}$ iteration loop. Replace $r$ by $r+1$ and repeat the cycle. Because this is no more than an implementation of the function representation that was established in Chapter Two, we know that the uniform convergence of $g_{q,r}$ and $f_r$ as $r \to \infty$ is guaranteed.

The iteration is shown schematically in Figure 5.22.

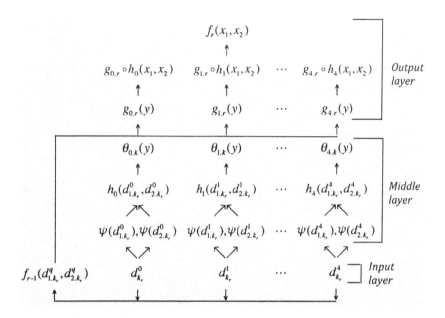

Figure 5.22: Schematic diagram of the first algorithm

Because $10 > 2 \cdot 4 + 1$, this algorithm can be used for functions of two, three and four variables. It has been a consistent observation that constructions and properties of representations of functions of two variables readily generalize to functions of arbitrary many variables. This remains true also to the computational algorithm above. It can be extended in two different ways: Using the base $\gamma \geqslant 2n + 1$ instead of 10, or using the universal function $\tilde{\psi}(x)$.

The first generalization follows verbatim by selecting a number $n > 2$ and using rational numbers

$$d_k^\gamma = \sum_{r=1}^{k} \frac{i_r}{\gamma^r}$$

for $i_1 = 0, 1, ..., \gamma$ and $i_r = 0, 1, ..., \gamma - 2$ otherwise; changes to conform to the new base are minor.

Once the universal function $\tilde{\psi}(x)$ has been constructed, its use in the representation of an arbitrary continuous function $f(x_1, ..., x_n)$

defined on $E^n$ follows the outline of the first algorithm, but a few changes should be noted. First and foremost, the choice of $n$ variables is followed by the selection of an integer $k_n \geqslant 2n + 1$. The function $f$ will be sampled on the Cartesian products

$$S_{q,k}(\tilde{d}_1, ..., \tilde{d}_n) = \prod_{p=1}^{n} \tilde{E}_{q,k}(\tilde{d}_{p,k}) = \prod_{p=1}^{n} [\tilde{d}_{q,k}, \tilde{d}_{p,k} + \delta_k], \quad k > k_n,$$

where $\tilde{d}_k = \sum_{r=1}^{k} \frac{i_r}{(r+4)!}$. A redefinition of the functions $\theta_{q,k}(y)$ will also be required, but with these adjustments and changes in notation and constants, the routine of the first algorithm can be followed.

Neither of these algorithms escapes entirely the dimensionality orbit because of the rapid increase in computational complexity in practice, even with more than a very small number of iterations. Rapid convergence at a low number of iterations, and alternative approaches, such as the Igelnik-Parikh cubic splines and others have shown, however, that function representation algorithms can be practical.

We know that an increase in the number of variables translates into a corresponding increase in the number of summand: computations are carried out in parallel, and added to this is the fact that the approximations are constructed on the real line.

From the perspective of computing, therefore, function representations can be interpreted as devices that enable the computation of a multi-variable function $f$ through $2n+1$ iterative parallel computations of which not less than $n+1$ approximate $f$ at any point of the domain. With available computing capacity, an increase from five to $2n+1$ parallel computations seems a relatively small prince to pay.

But there is a fly in this ointment. The algorithms developed above as well as constructions at various stages were carried out with only a few iterations that are in the nature of first approximations. It is evident that the implementation and proofs of convergence requite a target function $f(x_1, ..., x_n)$ to be known on a dense set of $E^n$. Specifically, the function must be known at points in any cube that intersects $E^n$ no matter how small.

For simplicity and efficiency we standardized the points at which $f$ is sampled to rational grid points $(d_1, ..., d_n)$ (to the base 10 or $\gamma$), but this may not always be appropriate: actual data may not include values $f(d_1, ..., d_n)$, or these may not be easily computed. In such cases, perhaps in many of applications, $f(x_1, ..., x_n)$ can be sampled at points that are interior to the cubes involved in the computation. The end result is clearly the same.

As we pointed out earlier, the functions $g_{q,r}$ that approximate $f$ could introduce large oscillations into the computation.

Key to Kolmogorov's function presentations is the strict monotonicity of the functions $\psi(x)$ that are the basis for transferring the computation of a function $f(x_1, ..., x_n)$ from an $n$-dimensional to a $(2n + 1)$-dimensional space. Recall that these functions served to map small $n$-dimensional cubes $S_{q,k}(d_{1,k}, ..., d_{n,k})$ of radius $\sqrt{n} \cdot 10^{-k}$ onto intervals $T_{q,k} = [\psi(d_{1,k}), ..., \psi(d_{n,k})]$. With the help of functions $\theta_{q,k}$ that equaled 1 on each of these intervals., local approximations $\theta_{q,k}(y) \cdot f_r(x_1, ..., x_n)$ were constructed in an iterative process. An alternative algorithm could be developed with interpolations of alternative curves passing through the grid-points $(d_{1,k}, ..., d_{n,k})$. These may result in approximating algorithms that do not converge to the superposition format, but could be an effective computational tool.

Any algorithm or method that we consider, however, requires knowledge of the linear ranking of grid-points $(d_{1,k}, ..., d_{n,k})$, and this confronts us with the particular space-filling curve that determines this ranking. This brings us squarely once again (no pun intended) to the computational Achilles heel that is inherent in these function representations.

The problem is inherent in the order of implementation:

$$d_k \rightarrow \psi_k(d_k) \rightarrow \Sigma_{p=1}^n \lambda_p \psi_k(d_{p.k}) \rightarrow g_k(y).$$

The functions $\psi(d_k)$ had to be constructed to satisfy the point-separating property described in Theorem 6, and this determined the a priori ranking of the grid-points $(d_{1,k}, ..., d_{n,k})$ that is at the heart of the problem. It was therefore imperative to optimize this ranking, but even the best possible Lipschitz continuity of functions $\psi(x)$ discovered by Fridman did not resolve this issue.

The realization that the a priori construction of $\psi(x)$ determined the ineffectiveness of any computational algorithm, raised the question of reversing the implementation: Is it possible to start with a space-filling curve having desirable computational characteristic, and then find a continuous function that it determines?

This conception follows the strategy that Kolmogorov used in his 1956 paper: Beginning with the universal tree $\Xi$, he constructed functions that led to function presentations. In analogy, we propose finding a suitable space-filling curve that would linearly rank the points of $E^n$, and finding continuous superpositions based on these for function representations.

The gain in computability is at the cost of the ultimate simplicity of Kolmogorov's formulation with functions of one variable and addition. Instead, we can achieve representations

$$f(x_1, ..., x_n) = \sum_{q=1}^{2n+1} g_q \circ \varphi_q(x_1, ..., x_n),$$

where the continuous functions $\varphi_q$ are independent of $f$, and the continuous functions $g_q$ that compute it are functions of one variable. By comparison, this is format of Kolmogorov's first representation formula in which the functions $\varphi_q$ are continuous functions with values in the universal tree.

Finding computationally optimal space-filling curve $s$ is an unexplored area. Computational algorithms have been developed for

two space-filling curve $s$. Detailed numerical tables and constructions are contained in (Sprecher [97], [98]).

An algorithm based on the Hilbert curve introduced geometrically in Chapter Three is easily obtained by emulating the first
algorithm. Reviewing Figure 4.11 reveals its easy geometric conception, but the actual numerical computation of the ranking of
its nodal points (their coding) requires some doing. A numbers
of papers contain different methods of coding the nodal points of
approximating curves and the reader is directed to (Sagan [82]) for
such detail. The author favors the method of (Chung, Huang, and
Liu [24]).

The Hilbert Curve is obtained by successively dividing the unit
interval $E$ into $2, 2^2, 2^3, 2^4, \ldots$ sub intervals, and thereby dividing
the square $E^2$ into $2^2, 2^4, 2^6, \ldots$ sub squares. In our application
we divide the interval $E$ into $8, 8^2, 8^3, 8^4, \ldots$ sub intervals, and the
square into $8^2, 8^4, 8^6, 8^8, \ldots$ subintervals. This amounts to skipping
the intermediate partitions. The reason for this deviation is that
we are emulating the construction of families of disjoint intervals
and squares that must diminish as a faster rate than by a factor of
2. We begin with the Hilbert space-filling curve $\mathfrak{H}_1$ in the square
$E^2 = [0, 1]^2$ with approximating curve $\zeta_1$ and make copies of $\mathfrak{H}_q$ in
the squares

$$\bar{E}_q^2 = \left[ -\frac{q-1}{7 \cdot 8}, 1 - \frac{q-1}{7 \cdot 8} \right]^2,$$

$q = 2, 3, 4, 5$; their intersection is the square[14]

$$\bar{E}^2 = \left[ 0, 1 - \frac{4}{7 \cdot 8} \right]^2.$$

---

[14]See Figure 5.23.

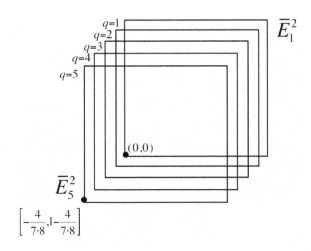

Figure 5.23: The squares $\bar{E}_q^2$.

The algorithm that we develop utilizes the Hilbert curve $\mathfrak{H}_1$ and its copies for the computation of functions $f(x_1, x_2)$ defined on $\bar{E}^2$ rather than $E^2$. This restriction is readily removed with the transformation:

$$x_p = \left(1 - \frac{4}{7 \cdot 8}\right) u_p,$$

and a renaming of the variables and functions. The algorithm is aimed at the following representation:

**Theorem 9.** *The Hilbert curve s $\mathfrak{H}_q$ determine real-valued continuous functions $\mathfrak{h}_q(x_1, x_2)$ defined on the square $\bar{E}^2$ such that every real-valued continuous function $f(x_1, x_2)$ with this domain has a representation*

$$f(x_1, x_2) = \sum_{q=1}^{5} g_q \circ \mathfrak{h}_q(x_1, x_2)$$

*with continuous functions $g_q$*

The functions $g_q$ that compute $f$ are functions of one variable, but lost is the ultimate simplicity because the functions $\mathfrak{h}_q$ cannot be expressed as a sum of continuous functions of one variable.

Let

$$\sigma_{q,k} = \frac{i_k}{8^k} - \frac{q-1}{7 \cdot 8} = \frac{i_k}{8^k} - (q-1)\sum_{r=2}^{\infty}\frac{1}{8^r}, \qquad (5.9)$$

for $i_k = 0, 1, ..., 8^k - 1$ and define the squares

$$\bar{S}_{q,k}^2(\sigma_{q,k}^1, \sigma_{q,k}^2) = \left[\sigma_{q,k}^1, \sigma_{q,k}^1 + \frac{6}{7 \cdot 8^k}\right] \times \left[\sigma_{q,k}^2, \sigma_{q,k}^2 + \frac{6}{7 \cdot 8^k}\right] \quad (5.10)$$

for $q = 1, 2, 3, 4, 5$ and $i_{p,k} = 0, 1, \ldots, 8^k - 1$. For fixed $q$, these are families of disjoint squares as shown in Figure 5.23 for $k = 1$. The squares coded by the Hilbert curve are designated $\bar{S}_{q,k}^2(\sigma_{q,k}^1, \sigma_{q,k}^2)_i$ as seen in Figure 5.24.

This explains the difference between the first algorithm and the one being developed here: In the algorithm based on Kolmogorov's representations, functions $\varphi(x_1, x_2) = \psi(x_1) + \lambda\psi(x_2)$ and their translates determine space-filling curve $s$ that determine the coding that linearly orders squares $\bar{S}_{q,k}^2$ for each $q$. We have seen the effect of this when we examined the anatomy of dimension reduction.

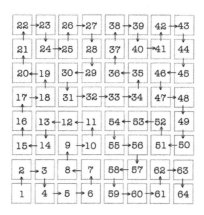

Figure 5.24: The coded squares $\bar{S}_{q,k}^2(\sigma_{q,k}^1, \sigma_{q,k}^2)$ for $k = 1$

The numerical lists are given in the following table:

| 1 | (0,0) | 9 | (2,2) | 17 | (0,4) | 25 | (2,6) |
|---|-------|----|-------|----|-------|----|-------|
| 2 | (0,1) | 10 | (3,2) | 18 | (1,4) | 26 | (2,7) |
| 3 | (1,1) | 11 | (3,3) | 19 | (1,5) | 27 | (3,7) |
| 4 | (1,0) | 12 | (4,3) | 20 | (0,5) | 28 | (3,6) |
| 5 | (2,0) | 13 | (1,3) | 21 | (0,6) | 29 | (3,5) |
| 6 | (3,0) | 14 | (1,2) | 22 | (0,7) | 30 | (2,5) |
| 7 | (3,1) | 15 | (0,2) | 23 | (1,7) | 31 | (2,4) |
| 8 | (2,1) | 16 | (0,3) | 24 | (1,6) | 32 | (3,4) |

| 33 | (4,4) | 41 | (6,6) | 49 | (7,3) | 57 | (5,1) |
|----|-------|----|-------|----|-------|----|-------|
| 34 | (5,4) | 42 | (6,7) | 50 | (7,2) | 58 | (4,1) |
| 35 | (5,5) | 43 | (7,7) | 51 | (6,2) | 59 | (4,0) |
| 36 | (4,5) | 44 | (7,6) | 52 | (6,3) | 60 | (5,0) |
| 37 | (6,5) | 45 | (7,5) | 53 | (5,3) | 61 | (6,0) |
| 38 | (7,5) | 46 | (6,5) | 54 | (4,3) | 62 | (6,1) |
| 39 | (7,6) | 47 | (6,4) | 55 | (4,2) | 63 | (7,1) |
| 40 | (5,6) | 48 | (7,4) | 56 | (5,2) | 64 | (7,0) |

To continue with the constructions: We associate with each square four unique corresponding squares,

$$\bar{S}^2_{q,k}(\sigma^1_{q,k}, \sigma^2_{q,k}),$$

for $q = 2, 3, 4, 5$.[15] The distances between their lower left hand vertices do not exceed $\frac{4\sqrt{2}}{7 \cdot 8^k}$, so that if a point $(x_1, x_2)$ belongs to any one of them, then

$$dist[(x_1, x_2), \sigma_{1,k}] \leqslant \frac{4\sqrt{2}}{7 \cdot 8^k}$$

Fix $q$ and define for each $k$ and $i$ the function

$$\alpha_{q,k}(x_1, x_2) = \sigma_{q,k}$$

when

$$(x_1, x_2) \in \bar{S}^2_{q,k}(\sigma^1_{q,k}, \sigma^2_{q,k}),$$

and set

$$\mathfrak{h}_{q,k}(x_1, x_2) = \sum_{i=0}^{8^{2k}-1} \alpha_{q,k}(x_1, x_2)_i$$

This function can be extended to a continuous piecewise planar function. The composition

$$f_1(x_1, x_2) = \sum_{q=1}^{5} g_{q,k} \circ \mathfrak{h}_{q,k}(x_1, x_2)$$

$$= \sum_{q=1}^{5} \sum_{i=0}^{8^{2k}-1} \alpha_{q,k}(x_1, x_2)_i \circ \frac{1}{3} f(\sigma^1_{q,k}, \sigma^2_{q,k})$$

follows in a natural way and this completes the first cycle of approximations to $f(x_1, x_2)$.

The algorithm is presented schematically in Figure 5.25 Once you get the hang of stringing the square $\bar{S}_{q,k}(\sigma^1_{q,k}, \sigma^2_{q,k})_i$ following the coding of the $k^{\text{th}}$ Hilbert approximating curves and superimposing on this chain one-third the function values $f(\sigma^1_{q,k}, \sigma^2_{q,k})_i$, the procedure becomes rote.

---

[15]See also Figure 5.23.

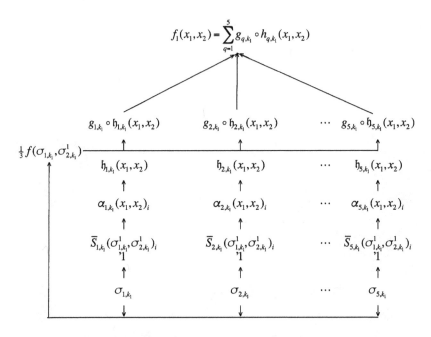

Figure 5.25: Second computational algorithm

These are the ingredients for this approximation algorithm. Select a number $\varepsilon < \frac{1}{8}$ and determine an integer $k_1$ such that
$$|f(x_1, x_2) - f(x_1', x_2')| \leqslant \varepsilon \|f\| \text{ when } |x_p - x_p'| \leqslant \frac{1}{8^k}$$

Let us suppose that we carried out the preceding steps for $k = k_1$. If $(x_1, x_2) \in \bar{E}^2$ is an arbitrary point, then by construction, there are least three values $q = \alpha_1, \alpha_2, \alpha_3$ such that $(x_1, x_2) \in \bar{S}^2_{\alpha_p k_1}(\sigma_{q,k_1}^1, \sigma_{q,k_1}^2)$, $p = 1, 2, 3$.

On each of these squares,

$$\left| \frac{1}{3}f(x_1, x_2) - \sum_{i=0}^{8^{2k}-1} \alpha_{q,k}(x_1, x_2) \circ \frac{1}{3}f(\sigma_{q,k}^1, \sigma_{q,k}^2) \right|$$

$$\leqslant \frac{1}{3} \left| f(x_1, x_2) - f(\sigma_{q,k}^1, \sigma_{q,k}^2) \right| \leqslant \frac{1}{3}\varepsilon \|f\|.$$

For the two remaining values of $q$:

$$\left| \frac{1}{3} f(x_1, x_2) - \sum_{i=0}^{8^{2k}-1} \alpha_{q,k}(x_1, x_2) \circ \frac{1}{3} f(\sigma_{q,k}^1, \sigma_{q,k}^2) \right| \leqslant \frac{1}{3} \|f\| .$$

Consequently,

$$|f(x_1, x_2) - f_1(x_1, x_2)| \leqslant 3\frac{1}{3}\varepsilon \|f\| + \frac{2}{3} \|f\| = \left( \varepsilon + \frac{2}{3} \right) \|f\| .$$

These estimates follow those in the proof of Kolmogorov's theorem in Chapter Two. Repeating the algorithm for $f(x_1, x_2) - f_1(x_1, x_2)$ instead of $f(x_1, x_2)$ results in a function $f_2(x_1, x_2)$ such that

$$|(f(x_1, x_2) - f_1(x_1, x_2)) - f_2(x_1, x_2)| \leqslant \left( \varepsilon + \frac{2}{3} \right) |f(x_1, x_2) - f_1(x_1, x_2)|$$

$$\leqslant \left( \varepsilon + \frac{2}{3} \right)^2 \|f\| .$$

An iteration leads to the general estimate

$$\left| (f(x_1, x_2) - \sum_{r=1}^{k} f_r(x_1, x_2) \right| \leqslant \left( \varepsilon + \frac{2}{3} \right)^k \|f\| .$$

We know that a uniformly convergent sequence of continuous functions converges to a continuous function. Without further explanation:

$$\|\mathfrak{h}_{q,k+m}(x_1, x_2) - \mathfrak{h}_{q,k}(x_1, x_2)\| =$$

$$\left\| \sum_{i=0}^{8^{2(k+m)}-1} \alpha_{q,k+m}(x_1, x_2)_i - \sum_{i=0}^{8^{2k}-1} \alpha_{q,k}(x_1, x_2)_i \right\| \leqslant \frac{1}{8^{k+m}}$$

since at any point $(x_1, x_2)$,

$$\left| \alpha_{q,k+m}(x_1, x_2)_i - \alpha_{q,k}(x_1, x_2)_i \right| \leqslant \frac{1}{8^m}.$$

Consequently, the sequence $\{\mathfrak{h}_{q,k}(x_1, x_2)\}$ converges uniformly for fixed $q$ to a continuous function $\mathfrak{h}_q(x_1, x_2)$.

The verification of convergence of the second sequence to a continuous function follows the same argument with the additional factor $\|f\|$. All this shows the algorithm converges to $f$.

In the above strategy, an a priori code determined by the Hilbert space-filling curve determines a function $\mathfrak{h}_{q,k}(x_1, x_2)$. We are able thereby control the linear ordering of squares $\bar{S}_{q,k}$ for each value of $q$, at the price of ultimate representations with only functions of one variable.

This opens the door to seeking alternative space-filling curve $s$ for the purpose of optimizing computations efficiency. Another such curve and an algorithm based on it can be found in (Sprecher [97]).

The conclusion drawn from the implementation of either dimension-reducing function representation model is that an underlying space-filling curve determined computational efficiency. Giving priority to this aspect rather than the simplicity of functions $\Sigma_{p=1}^{n} \lambda_p \psi(x_p + qa)$ may be a route worth exploring further.

## 5.5   Applications

The use of function representations in applications has proliferated since Hecht-Nielsen's interpretation of Kolmogorov's formula as a feedforward neural network, thereby shifting the paradigm from mathematical existence to computational algorithm. New applications cross the author's desk even as these lines are being written, and their a cataloguing is beyond the scope of this story. It seems appropriate, however, to mention a few salient examples. One of the telling tests often used for comparison of effectiveness is image compression, standardized through the use of the image of Lena that we already encountered. Mario Köppen applied a slightly modified version of our First Algorithm in a paper that corrected the numerical errors of the author noted earlier. A reconstruction of the Lena image for $k = 3$ was quite remarkable despite its granular appearance because of the single iteration that was used. It demonstrates again the rapid convergence that can be ascribed to the method of translations.[16]

Figure 5.26: Koppel's resampling of the Lena image
for a single value of $k$

Similar results were also achieved by Leni et all, who applied the first algorithm to bivariate functions, again verifying the rapid convergence that we noted. It was observed in this paper that *"preliminary work shows several interesting properties of the decomposition* [of target functions]. *An image can be converted into an ID signal: with bijectivity of function $\psi$, every pixel of the image is associated with a value in [0,1]."*

The authors gave voice to the implementation problems that

---

[16]See Figure 5.26.

we mentioned throughout the narrative; that in practice function representations have built-in hurdles with as many questions as answers. They asked:

> "How can space-filling curves be controlled? Can we generate different curves for some specific areas of the image? The main drawback of Sprecher's algorithm is its lack of flexibility:... space-filling curves, i. e. images scanning, is always the same function $[\psi]$."[63]

They point to the inevitable conclusion that the space-filling curve $s$ determined by the functions $\Sigma_{p=1}^{n}\lambda_p\psi(x_p + qa)$ determine the sweep of images being sampled. We have demonstrated that modifications subject to the point-separating conditions in Theorem 2.1 are of inconsequential effect on mitigating this computational Achilles Heel. The constants $\lambda_p$ can be adjusted, but again with seemingly little effect.

The question about generating alternative space-filling curve $s$ has been answered in the affirmative, but with the loss of format of Kolmogorov's function representations. Finding optimal curves is an open question.

This brings their criticism to the heart of the matter; namely, that the algorithm lacks flexibility. Hecht-Nielsen also raised these questions, as have others, and this remains an important area to explore. Of particular is the representation of functions based on uneven distributed sample data.

The Leni et al paper is particularly important because the authors pursued the reconstruction of the Lena image with two methods: the first algorithm as well as with the algorithm of Igelnik-Farakh using cubic splines [47].

This is an innovative architecture whose hallmark is flexibility. Using cubic spline $s$, his modified algorithm included degrees of adaptation to data that are lacking in the architecture based on Kolmogorov's function representations. This strategy was aimed at the approximate representation

$$f(x_1, ..., x_n) = \sum_{q=0}^{2n} g_q \circ \sum_{p=1}^{n} \lambda_p \psi_{pq}(x_p),$$

with the goal achieving a reduction in the complexity of computations and increasing its efficiency as a computational algorithm. The standard formula was replaced by

$$f_{igelnik}(x_1, ..., x_n) = \sum_{q=1}^{n} s \circ \left[ \sum_{p=1}^{n} \lambda_p s(x_p, \gamma^m), \gamma^n \right]$$

with positive constants $\lambda_p$ for which $\lambda_1 + ... + \lambda_n \leqslant 1$. This strategy replaced the functions $\psi_{pq}$ and $g_q$ with cubic spline functions. An important measure of this was the achieved degree of approximation:

$$\|f - f_{igelnik}\| = O\left(\frac{1}{N}\right)$$

that compares favorably with the bound $O\left(\frac{1}{\sqrt{N}}\right)$, where N is a predetermined positive integer.

This flexibility has its price: The inevitable casualty was the monotonicity of the transfer functions $\psi_{pq}$ that is essential to retain the superposition format in the limit, but the added flexibility of this technique makes function representations into an adaptable computational algorithm.

An interesting application is due to Valeriu Beiu, who used a computational algorithm for function representations to show that a minimum size neural network for implementing any Boolean function is obtained when the activation function of the neurons is the identity function [10].

An intriguing consequence is that the implementation requires analogue circuitry. There is an irony in this conclusion because, as we pointed out, Kolmogorov's function representations were analogue in their conception and formulation.

# Chapter 6

# Hilbert's Problem 13

## 6.1  Introduction

A capsule overview of this chapter was presented in Historical Synopsis, and readers whose sole interest is Kolmogorov's superpositions could stop here In doing so they would miss, however, an enriching detour that was the motivation for that discovery.

Algebra is an undeniable part of the complete story, specifically finding the roots of a polynomial equation – something that some may recall with disdain from their high school days, except that this time it was with a twist: Hilbert added continuous functions to the arithmetic operations and radicals that were the hallmarks of high school algebra.  In addition to leading to Kolmogorov's superpositions, one of the unforeseen consequences of this was the eventual connection with the exotic concepts of trees and space-filling curves.

The preface introduced side-by-side reduction in the number of coefficients of polynomial equations as a technique of computing their roots, and superpositions with continuous functions of one variable and addition. Some earlier material is revisited here from a purely algebraic perspective.

The narrative picks up here with one of the widely known algebraic functions: the roots of the quadratic equation

$$ax^2 + bx + c = 0, a \neq 0, \tag{6.1}$$

famously included in high school mathematics curriculum with the formulas:

$$\rho_1(a, b, c) = \frac{-b + \sqrt{b^2 - 4ac}}{2a} \quad \text{and} \tag{6.2}$$

$$\rho_2(a, b, c) = \frac{-b - \sqrt{b^2 - 4ac}}{2a}$$

The coefficients $a$, $b$, and $c$ are real numbers on which these roots depend, each choice giving a different answer, making the roots functions of these coefficients; of course $a \neq 0$. For example, the equation

$$x^2 - 4x + 4 = (x - 2)^2 = 0$$

a single root $x = 2$, whereas

$$x^2 - 4x + 3 = (x - 1)(x - 3) = 0$$

has two roots: $x = 1$ and $x = 3$.

As a general practice, rather than finding roots by brute force, as the above formulas do, we shall first simplify any equation that we wish to solve, and this will also simplify the formulas of the roots. Toward this end we use the the substitution

$$x = \left(\frac{c}{a}\right)^{1/2} \cdot u.$$

which simplifies equation (5.3) to the form

$$u^2 + Bu + 1 = 0 \tag{6.3}$$

where

$$B = \frac{b}{\sqrt{ac}}. \tag{6.4}$$

The formulas for the two roots of equation (6.2) are:

$$\sigma_1(B) = \frac{1}{2}B + \frac{1}{2}\sqrt{B^2 - 4} \quad \text{and} \qquad (6.5)$$

$$\sigma_2(B) = \frac{1}{2}B - \frac{1}{2}\sqrt{B^2 - 4}.$$

These roots depend on one coefficient only, and the roots of the original equation (6.2) can be computed using (6.4) and (6.5).

You could argue correctly that although formulas (6.5) are simpler than formulas (6.2), $B$ still depends on the three original coefficients, and so are implicitly the solutions of the equation (6.3), but we don't stop here because there is another point to be made.

For the next step, we write (6.4) in the form of a product:

$$\frac{b}{\sqrt{ac}} = (a^{-1/2} \cdot b) \cdot c^{-1/2}, \qquad (6.6)$$

and show how multiplication can be replaced by addition. For this, we recall from the Introduction the formula

$$x \cdot y = \frac{1}{4}(x + y)^2 - \frac{1}{4}(x - y)^2. \qquad (6.7)$$

An application of this formula to $a^{-1/2} \cdot b$ and then to $(a^{-1/2} \cdot b) \cdot c^{-1/2}$ gives after simplification:

$$B(a, b, c) = \frac{1}{4}\left\{\left[\frac{1}{4}(a^{-1/2} + b)^2 - (a^{-1/2} - b)^2 - (a^{-1} + b)\right] + c^{-1/2}\right\}^2$$

$$-\frac{1}{4}\left\{\left[\frac{1}{4}(a^{-1/2} + b)^2 - (a^{-1/2} - b)^2 - (a^{-1} + b)\right] - c^{-1/2}\right\}^2$$

This is a rather unwieldy expression for $\frac{b}{\sqrt{ac}}$, but there are two important point to it: The first is that the coefficients appear only in pairs or by themselves, in the case of $c$, and the second is that nowhere do you see multiplication. The squaring, which you can argue is a form of multiplication, is regarded here as an external function; that is $\frac{1}{4}(a^{-1/2} + b)^2 = \varphi(a^{-1/2} + b)$ as we already did.

We have here an example of a function of three variables, namely $B = \frac{b}{\sqrt{ac}}$ that is represented as a superposition of functions of one variable and addition, and we did this using only arithmetic. A lot of tedious computations that make a point!

The representation:

$$x \cdot y = h_1[a_{1,1}(x_1) + a_{1,2}(x_2)] + h_2[a_{2,1}(x_1) + a_{2,2}(x_2)]. \qquad (6.8)$$

is familiar from the Introduction as a lead-in to Kolmogorov's formula. These computations show that a variety of functions of any number of variables can be represent with arithmetic operations and radicals and a repeated application of expressions of the fixed form $h[a(x)+a(y)]$. In particular, all functions composed of the four arithmetic operations and radicals can be so represented, no matter how many variables are involved. This would include polynomial s and rational functions, and even functions involving radicals.

For example, the function

$$f(x, y, z) = \sum_{r=1}^{m} a_{i,j,k} x^{\alpha_i} y^{\beta_j} z^{\gamma_k},$$

in which the powers $\alpha_i, \beta_j, \gamma_k$ can be positive or negative numbers, can be so represented with a repeated application of formulas (6.7) and (6.8) by first applying it to the terms $(x^{\alpha_i} y^{\beta_j})$, and then to the terms $(x^{\alpha_i} y^{\beta_j}) z^{\gamma_k}$. In fact, looking through a calculus book the reader may be hard put to find functions where this is not possible.

With the tools that we just acquired we can obtain similar solutions for the general cubic equation:

$$ax^3 + bx^2 + cx + d = 0, a \neq 0. \qquad (6.9)$$

The transformation

$$x = \left(\frac{d}{a}\right)^{1/3} \cdot u$$

simplifies this equation to the equivalent form

$$x^3 + Ax^2 + Bx + 1 = 0,$$

in which we switched $x$ for $u$, and where

$$A = \frac{b}{a^{2/3}d^{1/3}}, \quad B = \frac{c}{a^{1/3}d^{2/3}}$$

after simplification.

The transformation $x = u - \frac{A}{3}$ leads to the following calculations:

$$x^3 = u^3 - Au^2 + \frac{A^3}{3} + \frac{A^2}{27}$$

$$Ax^2 = Au^2 - 2\frac{A^2}{3}u + \frac{A^3}{9}$$

$$Bx = BU - B\frac{A}{3}$$

Again changing notation by using $x$ instead of $u$, this cubic equation can be written in the form

$$x^3 + \alpha x + \beta = 0.$$

You already know the transformation for reducing it to

$$x^3 + \gamma x + 1 = 0$$

The simple transformation

$$u = x - \frac{\gamma}{3}$$

gives the equivalent equation

$$u^3 + \frac{\gamma^3}{27u^3} + 1 = 0,$$

and the substitution $u^3 = x$ gives after simplification

$$x^2 + x - \frac{C^3}{27} = 0$$

You know how to solve any quadratic equation, and we thus derived formulas for the roots of the general cubic equation (6.10) using only arithmetic operations and radicals.

It is natural to ask now if it is possible to keep going using the formulas for roots of lower degree polynomial equations to compute the roots of higher degree polynomial equations. This is possible once more with the quartic equation, but then our luck runs out: We can derive formulas for the roots of $x^4 + Ax + 1 = 0$, to which the general polynomial equation of the fourth degree $ax^4 + bx^3 + cx^2 + dx + 1 = 0$ can be transformed using the methods introduced earlier. What happens beyond that is where the story moves away from the familiar algebra and becomes more interesting and much more sophisticated.

Enter Évariste Galois, a French mathematician who died tragically in a duel at age twenty. In his short life span between 1811 and 1832 he made several contributions to mathematics, the most profound became known by the title *Galois Theory*

That is what anteceded our story: Galois established in the first third of the 1800s that general polynomial equations beyond the fourth degree are not solvable in closed form. This means that there are general formulas only for computing the roots of polynomial equations of degree 2, 3, and 4 by simply plugging in the specific coefficients, as long as they are real numbers. We have actually demonstrated this for degrees 2 and 3.

A further explanation is appropriate here: Its not that we cannot find formulas for the roots of higher degree polynomial equations; they simply don't exist! It's not a matter of trying harder and using better techniques. You might be able to find a formula for the roots of a specific polynomial equation of degree $n \geqslant 5$ or for a group of specific polynomial equations, but not in every case.

This is a detour on the way to Kolmogorov's superpositions from Hilbert's problem. Pursuing the reasons for this failure would carry the narrative too far from our goal by introducing concepts and a mathematical theory far beyond the scope of this narrative. Ii is important to note, however, that mathematics had to resort to alternative methods for computing roots, and this is where the story of Problem 13 begins.

## 6.2  What is problem 13 all about?

The introduction ended with the discovery that computing roots of polynomial equations beyond the fourth degree cannot be accomplished with formulas. This inability includes the roots of the specific seventh degree equation

$$x^7 + a_1 x^3 + a_2 x^2 + a_3 x + 1 = 0 \qquad (6.10)$$

that is at the heart of Problem 13. What attracted Hilbert's attention to this equation and how he came to select it from among all others is part of the story of this chapter.

The roots of an equation having two coefficients can be computed with nomographic methods, as can every function of two variables, but equation (6.10) has three coefficients! For a resolution, therefore, either nomographic techniques had to be extended to cover equations with three coefficients, or else one of the coefficients had to be eliminated. It turns out that it had to be the latter.

We already noted that for nomographic constructions, it is generally necessary that no more than two parameters be used at any given stage of the computation. As an example, in the equation $x^6 + ax^2 + bx + 1 = 0$ the parameters are the coefficients $a$ and $b$. The parameters of the equation

$$ax^6 + bx^2 + cx + 1 = 0, \qquad (6.11)$$

are $a$, $b$, and $c$, and to compute its roots with nomographic techniques we would

use the transformation $x = \frac{1}{a^{1/6}} u$ to obtain the equivalent equation

$$u^6 + Au^2 + Bu + 1 = 0 \qquad (6.12)$$

where $A = \frac{b}{a^{2/6}}$ and $B = \frac{c}{a^{1/6}}$. We now solve equation (1.12) with parameters $A$ and $B$ and then reverse substitution $u = a^{1/6}x$ to get the solution of equation (6.11).

Nomography was only the trigger for Problem 13, and we shall leave it at that. We know, however, that the roots of polynomial equations of degrees five and six – when regarded as functions of their coefficients, can be obtained through nomographic constructions because they involve only one and two coefficients, respectively as we show below. This gives us the first clue for Hilbert's selection of the seventh degree polynomial equation.

Later we present an excerpt of the text of Problem 13. It actually consists of two problems but the one that we were mostly concerned with stated that *the equation of the seventh degree*

$$x^7 + a_1 x^3 + a_2 x^2 + a_3 x + 1 = 0 \qquad (6.13)$$

*cannot be solved with the help of any continuous functions of only two variables.*

At first blush this looks like a problem in algebra: finding the roots of a very specific polynomial equation whose coefficients $a_1$, $a_2$ and $a_3$ are real numbers, but why bring functions into it? Several questions come immediately to mind: Why the concern with the roots of one particular polynomial equation? From the choice of any number of equations, why did Hilbert focus on one of the seventh degree? Why not eight, or any other number for that matter? The general equation of the seventh degree has eight arbitrary coefficients:

$$a_0 x^7 + a_1 x^6 + a_2 x^5 + a_3 x^4 + a_4 x^3 + a_5 x^2 + a_6 x + 1 = 0, \quad (6.14)$$

and we might wonder about the missing five terms:

But you remember what you learned in the introduction: preliminary to computing roots you simplify the polynomial equations. And speaking of solutions to a problem in algebra: algebra is algebra, and analysis is analysis, so how are continuous functions involved?

Yet what appears on the surface to be a mundane problem was presented by a rising-star to an august audience of mathematicians in a formal setting; it was included among the only ten problems that Hilbert included in his actual lecture. All this suggests caution in jumping to conclusions, especially in view of this excerpt from Hilbert's concluding remarks to his lecture:

> The problems mentioned are merely samples of problems, yet they will suffice to show how rich, how manifold and how extensive the mathematical science of today is...

There must have been therefore deep thought behind the choice of this specific problem, and plumbing this comes next, but first a rejoinder:

> We don't imagine even for a moment that the actual solution of this particular problem, if it could be found, was of sufficient importance to justify its inclusion in a set of twenty-three problems intended to chart the course of mathematics during the twentieth century.

As we already observed about problems in mathematics, the specific solution of a problem in and of itself is often of less importance than what it takes to get it. The interest lies instead in the window that the effort to find the solution opens for further discoveries, for moving mathematics forward; a problem's importance lies in the concepts, tools and theories that must be developed and refined in order to solve it, and Problem 13 was no exception.

We already eluded in the preface that this problem was the
harbinger or the seed, so to speak, of unintended and unimagined
developments that followed over five decades later—a short time-
span in a chronology of science but a long time on a human calendar.[1] The next section is devoted to examining this problem in
detail, and this means a long answer to the central questions that
we asked earlier: Why did Hilbert focus on a specific polynomial
equation of the seventh degree?

## 6.3   The Idea and the Method

The preceding pages were long on computations and short on answers, but we did learn what the questions were. We know now
that there is no formula for the roots of the general polynomial
equation of the seventh degree, or of any degree past four, for that
matter, but we really don't know why we should care. Mathematics offers many problems without solution, and why this particular
problem is of significance is as elusive as when we started.

We begin with the general polynomial equation of degree $n$:

$$a_0 x^n + a_1 x^{n-1} + a_2 x^{n-2} + ... + a_{n-1}x + a_n = 0 \qquad (6.15)$$

whose $n+1$ coefficients $(a_0, a_1, ..., a_n)$ stand for arbitrary real numbers.

We need to get rid of as many coefficients and terms as possible,
and the most obvious to eliminate is the constant term $a_n$, when it
is not zero. The equivalent equation

$$b_0 x^n + b_1 x^{n-1} + b_2 x^{n-2} + ... + b_{n-1}x + 1 = 0$$

is obtained with the substitution $b_r = \frac{a_r}{a_n}$, $r = 1, 2, ..., n$.

---

[1]See Lin [64].

The transformation $x = \frac{u}{b_0^{1/n}}$ results in the formula

$$u^n + c_1 x^{n-1} + c_2 x^{n-2} + \dots + c_{n-1} x + 1 = 0$$

in which $c_r = \frac{b_r}{b_0^{r/n}}$, and so with simple arithmetic we eliminated two coefficients. For subsequent purposes we want to relate all new coefficients to the original ones, and here $c_r = a_0^{-1/2} \cdot a_n^{-(n-1)/2} \cdot a_r$ after simplification.

The term $c_1 u^{n-1}$ can be eliminated all together with the transformation $v = u - \frac{c_1}{n}$ again using division to replace the constant term with 1 to obtain the equation:

$$v^n + d_1 v^{n-2} + d_2 v^{n-3} + \dots + d_{n-2} v + 1 = 0$$

It has $n$-2 coefficients and one less term. This transformation involves many computations that we omitted, but we note that only arithmetic operations and radicals were used in the process; we also note that the transformed equation has three less coefficients than the $n+1$ we started with. The new coefficients are composed, of course, of the original $n+1$ coefficients, but these would appear in the solution only after the fact, so to speak.

Changing notation, we see that we can find the roots of equation (6.15) by finding the roots of the equivalent equation

$$x^n + a_1 x^{n-2} + a_2 x^{n-3} + \dots + a_{n-2} x + 1 = 0$$

Less known today is the fact that also terms containing $x^{n-2}$ and $x^{n-3}$ can be eliminated with the help of the so-called Tschirnhaus Transformations, dating back to 1683 and generalizing the method of radicals.[2] Thus, using only arithmetic operations and radicals, the general polynomial equation (5.8) for $n = 5, 6, 7, 8, 9, \dots$ can be reduced to the normalized form:

$$x^n + a_1 x^{n-4} + a_2 x^{n-5} + \dots + a_{n-4} x + 1 = 0 \qquad (6.16)$$

---

[2]See Graver [39] and Wiman [111].

in which the number of coefficients was reduced from $n+1$ to $n-4$. Thus, the general polynomial equations of degrees 5 through 9 reduce to the normalized equations:

$$x^5 + a_1 x + 1 = 0$$

$$x^6 + a_1 x^2 + a_2 x + 1 = 0$$

$$x^7 + a_1 x^3 + a_2 x^2 + a_3 x + 1 = 0 \qquad (6.17)$$

$$x^8 + a_1 x^4 + a_2 x^3 + a_3 x^2 + a_4 x + 1 = 0$$

$$x^9 + a_1 x^5 + a_2 x^4 + a_3 x^3 + a_4 x^2 + a_5 x + 1 = 0$$

We find that to solve any polynomial equation, we need only solve its normalized form, and now we know why Hilbert used the specific seventh degree equation: The fifth and sixth degree equations can be solved using nomographic methods, and this is the first equation not so solvable.

We finally know why Hilbert chose this particular polynomial equation, but we don't know the significance of this choice. And the inclusion of functions in the conjecture is still as mysterious as ever, but as we stare at the list (6.17), a germ of an idea begins to crystalize and take shape in our mind.

It begins with the realization that a root $\rho(a_0, a_1, ..., a_n)$ of a general polynomial equation of degree $n$ depended on $n+1$ coefficients that are its variables: Each choice of coefficients gives different values to the root; in other words, the roots are functions of $n+1$ variables. This associated a polynomial equation of degree $n$ is with functions of $n+1$ variables.

But in practice the roots are functions of $n-5$ coefficients. True, they in turn are functions of the original $n+1$ coefficients, but only through superpositions using only arithmetic operations and radicals.

It occurs to us that this can be generalized into an important discovery: The roots of polynomial equations of degree $n$ as functions of $n+1$ variables can be represented as superpositions of continuous functions of $n$-5 variables. Roots are algebraic functions, and we arrived at the general statement that any algebraic function of $n \geqslant 5$ variables is representable as a superposition of algebraic functions of $n$-5 variables, and this is quite a discovery made with only arithmetic operations and radicals.

Reflecting on this further we also realize that the number of variables (coefficients) of a polynomial equation is not a useful criterion for classifying their roots as functions of the coefficient. The number of coefficients of normalized equations would be a better choice, but we don't know if this is as far as we can push the envelope: Is the normalized form in the list (6.17) for example the best possible, or are there other transformations that could further reduced the number of coefficients from equations of degree six onward?

This tantalizing thought brings us back to Problem 13. Now the question makes sense as far as the equation of the seventh degree is concerned: Is there a transformation that would eliminate one of the three remaining coefficients? This also explains using continuous functions for the purpose: Hilbert was persuaded that algebra would not do it, but this does not explain why he did not ask the same question of polynomial equations of degree higher than seven?

We view now Problem 13 in a new light: Five coefficients can be eliminated from general polynomial equations of degree $n \geqslant 5$, and we now ask if this is the best possible. We also have the story behind the conjecture that the equation of the seventh degree $x^7 + a_1 x^3 + a_2 x^2 + a_3 x + 1 = 0$ cannot be solved with the help of any continuous functions of only two variables.

# 6.4 Another Take on Problem 13

So now we know everything about Problem 13, or so it would appear. Although we did not read yet Hilbert's actual statement of the problem, we do know the story behind the conjecture, but we are left with lingering questions about the problem. The statement accompanying equation (6.13) was curt, and here is finally an actual excerpt:

> Now it is probable that the roots of the equation of the seventh degree is a function of its coefficients which does not belong to this class of functions capable of nomographic construction, i. e., that it cannot be constructed by a finite number of insertions of functions of two arguments.
>
> In order to prove this, the proof would be necessary that the equation of the seventh degree $f^7 + xf^3 + yf^2 + zf + 1 = 0$ is not solvable with the help of any continuous functions of only two arguments.[3]

The notation we used conforms to our adopted usage in this monograph; Hilbert's notation emphasizes that the focus is the roots $f(x, y, z)$ as functions of the coefficients are now the variables $x$, $y$, $z$, and this is an important distinction.

A side-by-side comparison of these two paragraphs suggests on first sight that they say the same thing in slightly different ways, but that in substance there is no difference between them; Namely, that the roots of the seventh degree polynomial equation cannot be expressed as a superposition of functions of two variables. But a closer reading shows that the term *continuous* is missing in the first paragraph, and being aware of the fussiness of mathematics when it comes to exact meaning makes us wonder if this was an editorial slip. The term *function* in mathematics is a broad umbrella and

---

[3]See Appendix A.

without qualification the first paragraph could be given different meanings.

The question of whether the second paragraph was a restatement of the first had arisen as soon as the problem became disseminated in the mathematical community. Consensus was quickly reached, however, that what Hilbert had in mind was a specific class of functions: that of analytic functions, and that he meant

> ...that it cannot be constructed by a finite number of insertions of analytic functions of two arguments.

This brings another concept into play, that of *analytic functions* that in the case of a single variable the reader had already encountered no doubt in calculus as a convergent power series:

$$f(x) = \sum_{r=0}^{\infty} a_r (x - c)^r$$

where the coefficients $a_r$ are real numbers, and the series converges in an open interval about the point $c$. Just as with polynomial s, the concept readily generalizes to several variables.

By now we have in our vocabulary three categories of functions: continuous functions, algebraic functions—the roots of polynomial equations, and now analytic functions as convergent power series. Each of these categories has its own characteristics, but we need not delve into this in any depth since we will not be developing these distinctions to a depth where these would be significant.

The above interpretation was also supported by the testimony of his contemporary Ludwig Bieberbach, but at any rate, the first paragraph above would be false as it stands, as Hilbert knew. It is easy to show, in fact, that any continuous function $f(x_1, ..., x_n)$ of $n$ variables can be represented as a sum of functions of only one variable.

The strategy for showing this is to partition the decimals

$$x_p = \sum_{r=0}^{\infty} \frac{i_{p,r}}{10^r} \quad , \quad 0 \leqslant i_{p,r} \leqslant 9.$$

into $n$ rows corresponding to $n$ coordinates, illustrated here for the case $n = 4$:

$$\frac{i_{1,1}}{10^1} + \frac{i_{1,2}}{10^5} + \frac{i_{1,3}}{10^9} + \frac{i_{1,4}}{10^{13}} + \frac{i_{1,5}}{10^{17}} + \frac{i_{1,6}}{10^{21}} + \cdots$$

$$\frac{i_{2,1}}{10^2} + \frac{i_{2,2}}{10^6} + \frac{i_{2,3}}{10^{10}} + \frac{i_{2,4}}{10^{14}} + \frac{i_{2,5}}{10^{18}} + \frac{i_{2,6}}{10^{22}} + \cdots$$

$$\frac{i_{3,1}}{10^3} + \frac{i_{3,2}}{10^7} + \frac{i_{3,3}}{10^{11}} + \frac{i_{3,4}}{10^{15}} + \frac{i_{3,5}}{10^{19}} + \frac{i_{3,6}}{10^{23}} + \cdots$$

$$\frac{i_{4,1}}{10^4} + \frac{i_{4,2}}{10^8} + \frac{i_{4,3}}{10^{12}} + \frac{i_{4,4}}{10^{16}} + \frac{i_{4,5}}{10^{20}} + \frac{i_{4,6}}{10^{24}} + \cdots$$

The pattern in which decimals are distributed on four coordinate lines is made clear by examining the powers in the denominators. To guarantee that each point $x_p$ has a unique representation, we stipulate that there is no integer $N$ for which $x_p = 9$ for all integers $r > N$.

Consider now the functions

$$h_p(x_p) = \sum_{r=1}^{\infty} \frac{i_{p,r}}{10^{4(r-1)+p}}, p = 1, 2, \ldots, n$$

They establish for each value of $p$ a one-one relation between the points of $E$ and a proper subset of. When $n = 4$ for example:

$$h_1(0.75) = 0.70005$$

$$h_2(0.75) = 0.070005$$

$$h_3(0.75) = 0.0070005$$

$$h_4(0.75) = 0.00070005$$

That the functions $h_p(x_p)$ are increasing with increasing decimals is self evident, and furthermore, $h_1(x_1) + \ldots + h_n(x_n) = h_1(x_1') +$

$... + h_n(x'_n)$ when, and only when$(x_1, ..., x_n) = (x'_1, ..., x'_n)$. The functions $h_p$ in this formula are discontinuous, and the one-one mapping $E \leftrightarrow E^n$ between the unit interval and the unit $n$-cube is also necessarily discontinuous.

We now know that each point $(x_1, ..., x_n) \in E$ has a unique image point $y = \sum_{p=1}^{n} h_p(x_p) \in E$ If we are given any function $f : E^n \to R$, then at every specific point $(x_1, ..., x_n) \in E$ we can define the function $g(y) = f(x_1, ..., x_n)$, giving the desired representation

$$f(x_1, ..., x_n) = g[h_1(x_1) + ... + h_n(x_n)]$$

It is important to understand that this is a pointwise definition.

The amended language makes the first paragraph into a deep mathematical problem. After everything that we have learned it may be an anticlimax to abandon our exploration of Problem 13 on this note, but we are not leaving it yet entirely. We still have much to learn from the views of mathematicians and computer scientists reflecting on this problem from a contemporary, post Kolmogorov superpositions vantage point that can be found in Appendix B. The point of the last example was to show why the first excerpted paragraph needed qualification.

The complete statement of Hilbert's problem can be found in Appendix A. In this connection, see Hilbert's Problems and their sequel (Kantor [50]).

# Appendix A

# Complete statement of Hilbert's Problem 13

13. Impossibility of the solution of the general equations of the $7^{\text{th}}$ degree by means of functions of only two arguments.

Nomography deals with the problem: to Solve equations by means of drawings of families of curves depending on an arbitrary parameter. It is seen at once that an equation whose coefficients depend upon only two parameters, that is every function of two independent variable, can be represented in manifold ways according to the principle lying at the foundation of nomography. Further, a large class of functions of three or more variables can evidently be represented by this principle alone without the use of variable elements, namely, all of those which can be generated by forming first a function of two arguments, then equating each of these arguments to a function of two arguments, next replacing each of those arguments in their turn by functions of two arguments, and so on, regarding as admissible any finite number of insertions of functions of two arguments.

So, for example, every rational function of any number of arguments belongs to this class of functions constructed by nomographic tables; for it can be generated by the process of addition, subtraction, multiplication and division and each of these processes produces a function of only two arguments. One sees easily that the roots of all equations which are solvable by radicals in the natural realm of rationality belong to this class of functions; for here the extraction of roots is adjoined to the four arithmetical operations and this indeed, represents a function of one argument only. Likewise the general equation of the 5$^{\text{th}}$ and 6$^{\text{th}}$ degrees are solvable by suitable nomographic tables; for, by means of Tschirenhausen transformations, which require only extraction of roots, they can be reduced to a form where the coefficients depend upon two parameters only.

Now it is probable that the roots of the equation of the seventh degree is a function of its coefficients which does not belong to this class of functions capable of nomographic construction, i. e., that it cannot be constructed by a finite number of insertions of functions of two arguments.

In order to prove this, the proof would be necessary that the equation of the seventh degree $f^7 + xf^3 + yf^2 + zf + 1 = 0$ is not solvable with the help of any continuous functions of only two arguments.

I may be allowed to add that I have satisfied myself by a rigorous process that there exist analytical functions of three arguments $x, y, z$ which cannot which cannot be obtained by a finite chain of functions of only two arguments.

By employing auxiliary movable elements, nomography succeeds in constructing functions of more than two arguments, as d'Octagne has recently proved in the case of the equation of the 7$^{\text{th}}$ degree**."

Hilbert cited two references:

* d'Ogtagne, Traité de Nomography, Paris, 1899.

** "Sur la tésolution nomographique de l'équation de septème degree." Comtes redus, Paris, 1900.

# Appendix B

# Comments on Problem 13

Hilbert formulated Problem 13 at the beginning of the last century in the context of the state of mathematics of that time. It was a nuanced problem as we have discovered, consisting as it does of two distinct problems: an algebraic problem that asks for superpositions of algebraic functions of three variables with analytic functions of two variables, and a conjecture that representations of algebraic functions of three variables with continuous functions of two variables was not possible. The first and deeper mathematically was the trigger for the second problem that, in turn, spurted the remarkable developments that will occupy our attention for most of this monograph.

As with any text, literary or scientific text, it is natural that it be reinterpreted by future generations looking at it through a long lens in the light of their own knowledge. Only few views preceding Kolmogorov's superposition theorem are found in the literature, so that the preponderance of view is retrospective, made in the reflected light of superpositions. They are enmeshed in that research and are an inseparable and important part of post- Kolmogorov developments.

The hybrid nature of the problem drew the attention of many contemporary mathematicians and computer scientists doing research on Kolmogorov's superpositions. Lorentz commented:

> ... it is much better to split the problem into two, an algebraic, and an analytic one

Arnold presented in a 1997 retrospective lecture a loose interpretation of Problem 13, saying that Hilbert asked in it if one really needs functions of three variables to solve the general seventh degree polynomial equation, and

> ... more generally, whether you can represent *any* function of three variables as a representation of functions of two variables – i.e., whether functions of three variables do exist.[5].

This echoes Lorentz's formulation. But following this general view, Arnold made as clear a statement as possible about the deeper meaning of the problem:

> Now I will discuss this problem returning to polynomials, and I will reformulate the Hilbert problem as it should be formulated. The function $z(a, b, c)$ that satisfies $z^7 + az^3 + bz^2 + cz + 1 = 0$ is an algebraic function of three variables. You can construct algebraic functions of three variables from algebraic functions of two variables by superpositions. The problem is whether this particular algebraic function $z(a, b, c)$ can be represented as a combination of algebraic functions of two variables. I would say that this was the *genuine* Hilbert problem. He did not formulate it in this way—unfortunately—and probably because he didn't, this problem is still open, and many times I have attempted to do something in this direction.

Kolmogorov himself expressed interest in the open part of Hilbert's problem, as shown in his letter.

Nowhere are these interpretations of Hilbert's intention made clearer than in an Irving Kaplansky unpublished chapter on Problem 13, part of an unfinished book on Hilbert's twenty-three problems. He goes as far as to imply that the Arnold - Kolmogorov's refutation: that every continuous function of three variables is a superposition of continuous functions of two variables, was beside the point, and he underscores this in the last two paragraphs of his draft chapter:

*"I am going to allow myself the liberty of some fantasy. What would Hilbert have said if the Kolmogorov - Arnold theorem had been presented to him? After the initial surprise and delight, it seems to me he would have remarked that the function in which he was interested (the roots of the normalized 7-th degree equation) is an* algebraic *function. Therefore, it would be more reasonable to insist that only algebraic functions be used in the representation. In this form the problem does not seem to have been studies (although Vituškin mentioned it).*[1]

He dismisses entirely that Hilbert may have had any interest in the purely analytic problem, and he may have well been right in that. Hilbert's work on the 9th degree polynomial equation, to which we return many years later, tended to support Kaplansky's flight of fantasy. The first published attempt at a partial solution was due to Bieberbach. He had published four papers on Hilbert's problems: about nomography in 1922 [11]; the impact of the problems in 1930 [12]; an ill-fated attempt at solution to Problem 13 in 1931 [13]; and a correcting note was published in 1934 [14].

He presented the representation:

$$f(x_1, x_2, x_3) = F[x_1, \varphi(x_2, x_3)]$$

for an arbitrary function $f(x_1, x_2, x_3)$, noting that such representations involve discontinuous functions, as we have seen in the example above.

---

[1]See Appendix C.

The key result that he was after was that not every continuous of three variables can be represented with superpositions of continuous functions of two variables.

# Appendix C

# Irving Kaplansky's Draft

Some time in the 1980s—I regret not having a record of the date—Irving Kaplansky contacted me to say that he was writing a monograph on Hilbert's twenty three problems. He sent me a draft of chapter 13 for comment, and we had one or two telephone conversations and a brief correspondence on the subject that is regretfully lost. The draft contains a number of minor inaccuracies that are noted below, perhaps attributable to 13 being an unlucky number; Hilbert and Bieberbach as well as the author did not escape. Neither Kaplansky nor I followed up on this initial contact. This is now part of a set of lecture notes at the University of Chicago.

I still have in my files his hand-corrected draft, and reviewing it among other materials when writing this monograph I was struck, as I was at the time, by the elegance of Kaplansky's style, the economy of the presentation, and the emphasis that he placed on the algebraic perspective that was inherent in Hilbert's thinking, but stripped of it since Kolmogorov's paper moved Problem 13 into the realm of mathematical analysis. The representation is a fairly complete insightful sketch of the problem, except for two references and missing in his extensive bibliography; the draft says little about the extensive work surrounding Kolmogorov's theorem.

Even though there is considerable overlap between Kaplansky's draft and the body of the text, I thought it proper to include it verbatim in this appendix. The only exception is the bibliography that is subsumed in the expanded listings of this monograph. As a result, Kaplansky's bibliographical references bear different numbers.

To obtain permission for including the draft in this monograph, I turned to Internet resources as a matter of course. Googling Kaplansky's name led to the website of his singer and songwriter daughter, Lucy Kaplansky. I am in her debt for the lightening speed and gracious responses shown in the following e-mail exchanges:

April 20, 2015 11:38 AM

To: Lucy Kaplansky - Permission to use material of Irving Kaplansky

My name is David Sprecher, retired mathematician writing a monograph on a subject that was of late interest to your father. About 15[1] years ago he sent me a draft chapter that he was working on and never completed (as far as I know). It is well written and an interesting perspective on the subject of my monograph (on Hilbert's Problem 13). I would like to add his insightful chapter as an appendix, and ask for your permission to do so. With all best regards, David Sprecher Professor Emeritus

April 21, 2015 9:11 AM

Hi David

Thank you so much for the email!

I've forwarded your email to my brothers and I'm just waiting to hear back from them. I'll email you when I do.

All the best, Lucy

---

[1]This time reference was in error.

April 21, 2015 10:51 AM

Dear Lucile,

Thank you most profusely for your quick response to my inquiry. My area of research in mathematics was generally far removed from those of your father, and we overlapped only on one area. Having been an admirer of his (among so many) I want to honor his memory with the inclusion of an unfinished piece of work.

On a personal note, are you still performing in New York? I live in Manhattan and would love to see you on stage, as would my family.

Thanks again,

David

April 21, 2015 12:44 PM

You're so welcome!

I just heard from my brothers and we're all totally fine with it. Thank you so much for checking with us.

I do perform in NYC, once or twice a year. Best thing is to check my website every now and then for shows, I don't have anything booked right now.

Thanks again.

All the best, Lucy

Chapter 13. Impossibility of the solution of the general equation of the 7-th degree by means of functions of only two arguments.[2]

Irving Kaplansky

(First Draft)

There were two sources for Hilbert's interest in the ideas he put forward in the thirteenth problem.

In 1886, shortly after his doctorate, Hilbert visited Paris. It is noted in Reid's biography [81, pager 23-24] that he became quite friendly with D'Octagne. Thus it was natural for him to follow with interest D'Ocagne's work which culminated in his book [25] on nomography, the definitive treatise on the subject at that time.

The other topic to attract Hilbert's attention was the venerable one of endeavoring to put polynomial equations (in one variable) in as simple a form as possible.

Nomography studies schemes for graphical solutions of equations. The Britannica article [34] by Ford could serve as a good introduction for a reader unfamiliar with the ideas. Another reference is the survey article [11] by Bieberbach. We shall not attempt an exposition here, and shall just say that the theory does suggest that it would be nice to have a way of expressing functions in a certain number of variables in terms of functions of fewer variables. Hilbert emphasizes the case where the number of variables is to be reduced from 3 to 2.

Such a reduction might, for example, look as follows: we are given a function $f$ of 3 variables, and functions $g_1, g_2, g_3$ of two variables in such a way that

$$f(w, y, z) = g_1[g_2(x, y), g_3(x, z)].$$

---

[2]Bibliographic references conform to the bibliography of this monographs. Items in Kaplansky's bibliography that were not referred to in the text have been omitted.

Consider now the question of simplifying the general polynomial equation of degree $n$ in one variable:

$$x^n + a_1 x^{n-1} + a_2 x^{n-2} + ... + a_{n-1} x + a_n = 0.$$

The first thing we do with this equation is to get rid of the coefficient of $x^{n-1}$ by the transformation $x = y - a_1/n$. (To be cautious, this requires a characteristic prime to $n$.) Not so well known these days is the fact that by a transformation using only extraction of roots (in addition to rational functions) it is possible to remove the next two coefficients, $a_2$ and $a_3$. The transformation accomplishing this is named for Tschirnhaus, who published it in 1683. One account of the Tschirnhaus transform can be found on pages 95–98 of [G. Bauer. *Vorläsungen über Algebra*, Teubner, 1003. Revised by L. Bieberbach, Taubner, 1928, pages 122–124 of Bieberbach's revision]. This theory is of course a generalization of the solution by radicals of equations of degree $\geq 4$.

By a further extraction of an $n$-th root we can make the constant term 1. After this the equation of the 5-th degree takes the form

$$x^5 + ax + 1 = 0. \qquad (*)$$

Thus equations of the 5-th degree can be solved by rational operations, extraction of roots, and the one further function of one variable given by the solution of $(*)$ as a function of $a$.

Likewise the normalized equation of the 6-th degree

$$x^6 + ax^2 + bx + 1 = 0$$

yields a function of two variables, and we might wonder (as Hilbert did in [45]) whether a reduction to functions of one variable is possible.[3]

In his 1930 survey of the status of Hilbert's problems [12], Bieberbach said that Hilbert meant that his analytic functions of

---

[3]See excerpt from Hilbert's paper on page 196.

three arguments could not be expressed in terms of analytic functions of two arguments. Since this correction was made during Hilbert's lifetime, it is surely authentic.

The first published result of this kind appears to be the one due to Ostrowski [77], mentioned by Hilbert in [45]. This asserts that the analytic function of two variables

$$\zeta(u, v) = u + \frac{u^2}{2^v} + \frac{u^3}{3^v} + \dots + \frac{u^n}{n^v} + \dots$$

cannot be expressed in terms of rational functions and analytic functions of one variable Further references are [A. G. Vituškin. Proof of existence of analytic functions of several variables not representable by linear superpositions continuously differentiable functions of fewer variables, Doklady Akad. Naul 157(1964), 1258-1261. (Russian)] and the opening portion of Arnold's survey [3].

But if we allow any continuous functions of fewer arguments to be used, then Hilbert was wrong in his conjecture that the number of variables cannot be depressed in the function arising from the equation of the 7-th degree. In fact, any continuous function of $n$ variables can be represented by the use of addition and continuous functions of just one variable![4]

The breakthrough in proving this came in the paper [54] by Kolmogorov. In it he showed that, for any $n$, a continuous function of $n$ variables can be expressed in terms of continuous functions of 3 variables.[5] Of course, this was not quite good enough to settle the 13-th problem the way Hilbert stated it. But shortly thereafter, Kolmogorov's student Arnold [1] reduced 3 to 2 and thus achieved the negative solution of Hilbert's problem. In [55] Kolmogorov proved the still stronger result mentioned above, that is, he was able to get by with continuous functions of one variable and just one function of two variables: addition. At the same time he simplified

---

[4]The functions must be uniformly continuous.
[5]Ditto.

the proof. The actual theorem proved is even more remarkable in that all but one of the functions used can be fixed in advance. For simplicity we state it for; the changes needed for arbitrary $n$ are nominal (for instance, $2n+1$ $\varphi's$ are used in place of 5).

**Theorem 10.** *Fix a real number $\lambda$ with $0 < \lambda < 1$. There exist five real continuous strictly monotone functions $\varphi_1, ..., \varphi_5$, defined on [0,1] and taking values in [0,1], with the following property: for any continuous real function $f(x,y)$ defined for $0 \leqslant x \leqslant 1$, $0 \leqslant y \leqslant 1$, there exist a continuous real function $g(u)$ defined for $0 \leqslant u \leqslant 2$, with[6]*

$$f(x,y) = \sum g[\varphi_i(x) + \lambda\varphi_i(y)]^3.$$

The proof is quite elementary; the most advanced result used is the uniform continuity of a continuous function on a compact (i.e. bounded closed) subset of Euclidean space. But it is nevertheless intricate and requires sustained attention on part of the reader. Roughly speaking, the procedure is to subdivide in an appropriate way the unit interval and the unit cube in $n$-space, and then devise a way of "packing" the latter into the former, due account being taken of the given functions of $n$ variables. Then the subdivision is taken ever finer, and in the limit the desired representation is obtained.

An excellent account of the Kolmogorov-Arnold theorem appears on pages 168-174 of Lorentz's book [67].

There are two theorems of related work in the literature. There are affirmative results of the kind discussed above for analytic functions. The pattern appearing in these theorems is as follows: for various notions of smoothness stronger than mere continuity it is not possible to write functions of $n$ variables in terms of functions of $n$-1 variables; it is necessary to relax the smoothness requirements in order to depress the number of variables. Some references

---

[6]The summation over $i$ is from 1 to 5.

are: the papers of Vituškin in the bibliography, Arnold's survey [4], pages 174-178 of Lorentz's book [67], and chapter 13 of the Russian book (this chapter was written by Vituškin). In the second stream of work the Kolmogorov-Arnold theorem has been sharpened in various ways. See the papers of Doss, Ostrand, and Sprecher in the bibliography.

I am going to allow myself the liberty of some fantasy. What would Hilbert have said if the Kolmogorov-Arnold theorem had been presented to him? After the initial surprise and delight, it seems to me he would have remarked that the function in which he was interested (the roots of the normalized 7-th degree equation) is an algebraic function. Therefore, it would be more reasonable to insist that only algebraic functions be used in the representation. In this form the problem does not seem to have been studies (although Vitŭskin mentioned it). Incidentally, Hilbert attacked the normalized equation of the 9-th degree in [45]; this is an algebraic function of five variables, and when he proved that it was expressible in terms of functions of four variables, he did indeed use algebraic functions. For more on this, see papers of Garver, Tschebotaröw, and Weiman listed in the bibliography. To conclude this chapter, I shall record a way of recasting the problem just mentioned. When algebraic functions are treated analytically, the resulting multiple valued functions are awkward. Also an algebraist is unhappy with the restriction to the filed of complex numbers. Let us make it a piece of algebra, In the simplest case, the problem takes the following form.

Let $k$ be an algebraically closed filed. Let $u$, $v$ be indeterminates over $k$, and let L be the algebraic closure of $k(u,v)$. Let H be a field lying between $k(u,v)$ and L and enjoying the following property: along with any element z, H contains the algebraic closure of $k(z)$. Is H necessarily equal to L? Does H contain $f$, where $f^6 + uf^2 + vf + 1 = 0$?

# Appendix D

# End Matter

## D.1  Kolmogorov's letter

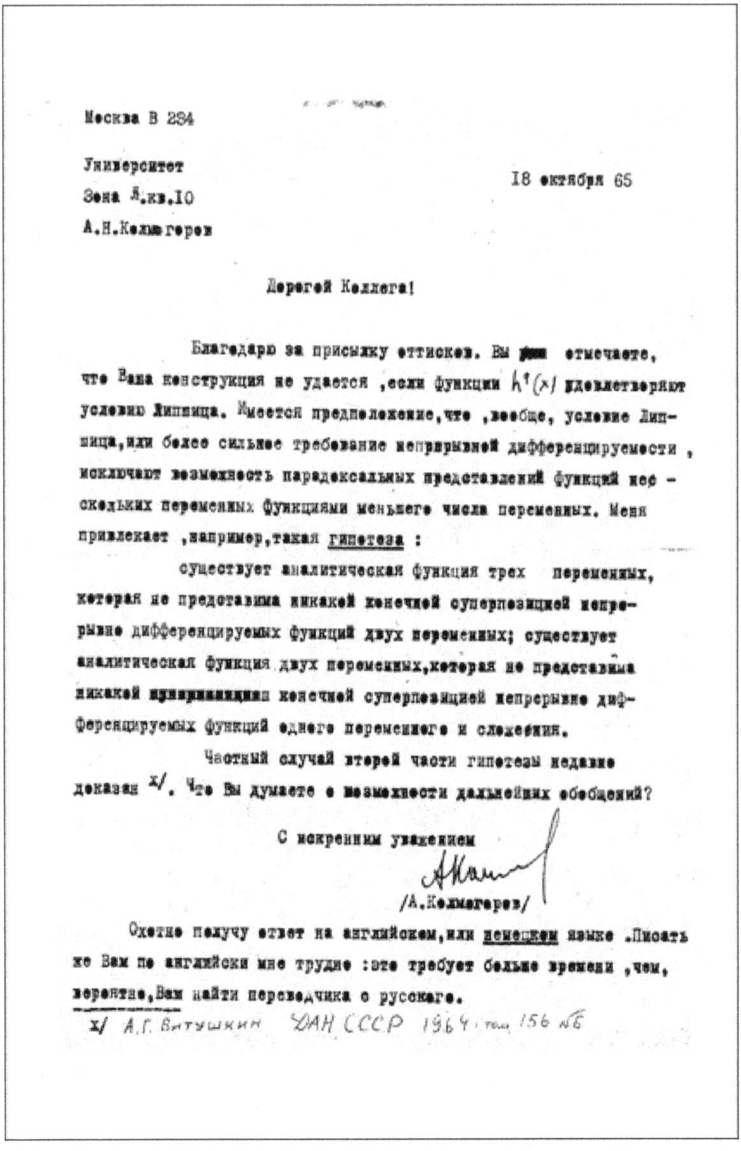

Москва В 234

Университет
Зона А.кв.10                                    18 октября 65
А.Н.Колмогоров

                    Дорогой Коллега!

        Благодарю за присылку оттисков. Вы отмечаете,
что Ваша конструкция не удается ,если функции $h^1(x)$ удовлетворяют
условию Липшица. Имеется предположение,что ,вообще, условие Липшица,или более сильное требование непрерывной дифференцируемости ,
исключают возможность парадоксальных представлений функций нес-
кольких переменных функциями меньшего числа переменных. Меня
привлекает ,например,такая гипотеза :
        существует аналитическая функция трех   переменных,
которая не представима никакой конечной суперпозицией непрерывно дифференцируемых функций двух переменных; существует
аналитическая функция двух переменных,которая не представима
никакой конечной суперпозицией непрерывно дифференцируемых функций одного переменного и сложения.
        Частный случай второй части гипотезы недавно
доказан $^{x/}$. Что Вы думаете о возможности дальнейших обобщений?

                    С искренним уважением

                            /А.Колмогоров/

    Охотно получу ответ на английском,или немецком языке .Писать
же Вам по английски мне трудно :это требует больше времени ,чем,
вероятно,Вам найти переводчика с русского.
    ————————
    x/ А.Г. Витушкин  ДАН СССР 1964 том. 156 №5

Figure D.1: Kolmogorov's letter

Moscow University B234 October 10, 65
Zone l.kv.10
A.N. Kolmogorov

Dear colleague!

Thank you for the sent imprint. You have already noticed that your construction is not successful if h $^9(x)$ function satisfies a Lipschitz condition. There is a hypothesis that in general the Lipschitz condition or stronger requirement of continuous differentiability excludes possibility of antinomous representation of functions of several variables by less number of variables. I personally like the following hypothesis: There exists analytical function of three variables that could not be represented by any finite superposition of continuously differentiable function of two variables. There exists analytical function of two variables that is not represented by any finite superposition of continuously differentiable functions of one variable and addition. The special case of the hypothesis second part was recently proven x/. What do you think about possibility of the future synthesis? Best regards,

A. Kolmogorov

I will be happy to receive your answer in English or German. It is hard for me to write in English — this requires more time then to find a translator from Russian for you.

x/ A.G. Vitushkin DAN SSSR 1964 Vol. 156 No. 6

## D.2   Vituškin's lecture notes

The text of this undated lecture note of 1964 agrees in substance with the abstract [200]. In the note "Gilbert" is " Hilbert, "tryed" is "tried," and "Chenkin" is " Henkin."

VITUSHKIN A.G. ON THE POSSIBILITY OF
FUNCTION REPRESENTATION WITH SUPERPOSITIONS OF
FUNCTIONS WHICH HAVE LESS VARIABLES

With the help of algebraic substitution
named Chirngauzen transformation the common
algebraic equation to the $n$ power $x^n + a_1 x^{n-1} + a_2 x^{n-2} + \ldots + a_n = 0$ is reduced to the form
$y^n + b_4 y^{n-4} - b_5 y^{n-5} \ldots + b_{n-1} y + 1 = 0.$ Further attempts of
the algebraists to reduce the solution of common
algebraic equation to the solution of the equation
with the less number of parameters ( if possible)
remained unsuccessful for a long time.

In his "Mathematical problems" D.Gilbert
approached this task in a new way. He formulated
it by N 13 as follows: "the impossibility of the
solution of a common equation to the power 7 with
functions of only two variables". To prove it
G.Gilbert considered it neccessary to prove that
"the equation to the power 7 $f^7 + x f^3 + y f^2 + x f + 1 = 0$
couldn't be solved by means of any continuous
functions of only two variables".

Mathematicians have understood the 13-th
problem differently and attributed to it the results of different character.

A.Ostrovsky showed (1920) that the analytical
function of two variables $\zeta(x,y) = \sum_{n=1}^{\infty} \frac{x^n}{n^y}$ is not
a finite superposition of infinitely differentiated
functions of one variable and algebraic functions

Figure D.2: Vituškin Lecture Notes

of any number of variables. D.Gilbert (1926) proved that the solution of an equation to the 9 power can be represented as a finite superposition of 4 variables algebraic functions.

In 1955 the author proved that there exists a $\rho$-time differentiated function of $n$-variables, which cannot be represented as a finite superposition of $\rho'$-time differentiated functions of $h'$ variables if only $n/\rho > n'/\rho'$

L.Bieberbach (1929) tryed to prove that there exists continuous function of three variables, which cannot be represented as a finite superposition of continuous functions of two variables. Really, it was not for nothing that L.Bieberbach called the 13-th problem unfortunate.

In 1957 the joint efforts of A.N.Kolmogoroff and V.I.Arnold resulted in proving the reverse: any continuous function of $h$ - variables can be represented with such a superposition as $\sum_{i=1}^{2n+1} f_i \left( \sum_{j=1}^{n} \alpha_{ij}(x_j) \right)$ where all the functions are continuous but inner functions $\{ \alpha_{ij}(x_j) \}$ are previously fixed.

From this theorem follows, in particular, that the solution of the equation to the power 7 can be represented as the superposition of the continuous function of one variable and an addition operation. So Gilbert's hypothesis proved to be faulty.

In connection with the theory of Fourier series
N.K.Bari proved as far as in 1930 that any continuous function of one variable can be represented
as $f(x) = \sum_{i=1}^{\infty} f_i(g_i(x))$ where all $\{f_i\}$ and
$\{g_i\}$ are absolutely continuous. That is why any
continuous function of $h$ - variables can be represented as the superposition with absolutely continuous function of one variable and an addition
operation.

In the recent years a number of results supp-
lementing Kolmogoroff's theorem was received (L.A.
Bassalyga, G.M.Chenkin, R.Doss, G.G.Lorentz, P.A.
Ostrand, D.Sprecher, V.M.Tichomiroff a.o.).

While considering the superpositions of differentiated functions the character of the statements
seems to be essentially changed.

For example, it follows from the papers of the
author and G.M.Chenkin that for any set of continuous functions $P_m(x_1, x_2, \dots x_n)$ and continuously
differentiated functions $q_m(x_1, x_2, \dots x_n)$, $n > 2$, $(m = 1, 2, \dots N$,
a set of superpositions $\sum_{i=1}^{N} P_i(x_1, x_2 \dots x_n) \cdot f_i(q_i(x_1, x_2 \dots x_n))$
where $\{f_i(s)\}$ are arbitrary continuous
functions of one variable, is nowhere dense in
continous function space of $h$ - variables.

In conclusion it would be neccessary to mention
that Gilbert's idea to prove "the impossibility of
the solution of any common equation to the power 7 by means of the function of only two vari-

ables" may receive further positive development.

All the results known up till now don't contradict, for example, the hypothesis that the function $f(x,y,z)$ defined by the equation $f^7 + x f^3 + y f^2 + x f + 1 = 0$ is not a finite superposition of analytical functions of two variables

# Bibliography

1. V. I. Arnold. *On functions of three variables.* Dokl. Akad. Nauk SSSR 114, 679-681. Transl. Amer. Math. Soc. 2(28)1963, 51-54, 1957.

2. V. I. Arnold, V.I. *On the representability of functions of two variables in the form F[G(X)+H(Y)].* Uspehi Mat. Nauk (N.S.) 12(2)74, 119-121, 1957.

3. V. I. Arnold. *Some questions on approximation and representation of functions.* Proc. International Congress of Math., 339-348, Cambridge Univ. Press, New York, 1960. Transl. Amer. Math. Soc. 2(53)1966, 192-201, 1958.

4. V. I. Arnold. *On the representation of continuous functions of three variables by superpositions of continuous functions of two variables.* Mat. Sb. (N.S.) 48(90), 3-74, Transl. Amer. Math. Soc. 2(28)1963, 61-147, 1959.

5. V. I. Arnold. *From Hilbert's superposition to Dynamical Systems.* Math. Assoc. Amer. Monthly 111(7), 608-624, 2004.

6. S. Banach. *Theory des Opérations Linéaires.* Chelsea Publishing Compay, New York,

7. Dor Bar-Nathan. *Dessert: Hilbert's 13$^{th}$ Problem, in Full Colour.* Dept. of Math. Univ. of Toronto, 2009.

8. Nina Bari. *Memoire sur la representation finie des fonctions continues.* Math. Ann. 103, 145-248, 598-653, 1930.

9. L. A. Bassalygo. *On the representation of continuous functions of two variables by means of continuous functions of one variable.* (Russian, English summary) Vestnik Moskow. Univ. Ser. I Mat. Meh. 21(1), 58-63, 1966.

10. V. Beiu. *On Kolmogorov's Superpositions and Boolean Functions.* Los Alamos National Laboratory 1998.

11. L. Bieberbach. *Über Nomographie. Die Naturwisswnschaften* 10, 775-782, 1922.

12. L. Bieberbach. *Über den Einfluss von Hilberts Pariser Vortrag über "Mathematische Probleme" auf die Entwicklung der Mathematk in den letzten dreissig Jahren. Die Naturwisswnschaften* 18, 1101-1111, 1930.

13. L. Bieberbach. *Bemerkunen zum dreizehnten Hilbertschen Problem, J. Reine Angew. Math.* 165, 89-92, 1931.

14. L. Bieberbach. *Zusatz zu meiner arbeit "Bemerkungen etc." in Band 165 dieses Journals, J. Reine Angew. Math.* 170, 242, 1934.

15. E. K. Blum, L. K. Li. *Approximation theory and feedforward neural networks.* Neural Networks 4, 511-515, 1991.

16. S. Bochner. *Lectures on Fourier Integrals.* Princeton Univ. Press, Princeton, N.J. Chapter VI, 1959.

17. V. Brattka. *From Hilbert's 13$^{th}$ Problem to the theory of neural networks: constructive aspects of Kolmogorov's Superposition Theorem.* Kolmogorov's Heritage in Mathematics, 253-80, Springer Verlag, 2007.

18. V. Brattka. *A Computable Kolmogorov Superposition Theorem*. Theoretische Informatik I, Fern Universität Hagen, German

19. J. Braun. *Kolmogorov's Superposition Theorem to Function Reconstruction in Higher Dimensions*. Dissertation, Rheinischen Friedrich-Wilhems Univ., Bonn, 2009.

20. R. C. Buck. *Approximate functional complexity*. Bul. Amer. Math. Soc. 81(6), 112-114, 1975.

21. R. C. Buck. *Approximate functional complexity*. Approximation Theory II, Academic Press, N.Y., 303-307, 1976.

22. R. C. Buck. *Approximate complexity and functional representation*. J. Math. Analysis and Applications. 70, 280-298, 1979.

23. H. Chen, T. Chen, R. Liu. *A constructive proof of approximation by superposition of sigmoidal functions for neural networks*. Preprint, 1-6, (1993).

24. K. l. Chung, Y. L. Huang, and T. W. Liu. *Efficient algorithms for coding Hilbert curve of arbitrary-sized image and application to window query*. Information Scienc 177(2007), 2130-2151.

25. Maurice d'Ogcagne. *Traiťe de Nomographie*. Paris. 1899

26. Maurice d'Occagne, *Sur la resolution nomographique de l'équation du septiéme degree*. *C. R. Paris* 1315, 22-24, 1900.

27. S. P. Diliberto, E. G. Straus. *On the approximation of a function of several variables by the sum of functions of fewer variables*. Pacific J. Math, 1, 195-210, 1951.

28. Rauf Doss. *On the representation of continuous functions of two variables by means of addition and continuous Functions of one variable.* Colloquium Math. 10(2), 249-259, 1963.

29. Rauf Doss. *Representation of continuous functions of several variables.* Amer. J. Math. 98, 375-383, 1976.

30. Rauf Doss. *A superposition theorem for unbounded continuous functions.* Trans. Amer. Math. Soc. 233, 197-203, 1977.

31. Epstein. *Nomography.* Interscience, New York 1958.

32. H. A. Evesham. *The History and Development of Nomography.* Docent Press, Boston, 1982.

33. R. J. P. de Figueiredo. *Implications and applications of Kolmogorov's superposition theorem.* IEEE Trans. Automatic Control AC-25, 1227-31, 1980.

34. Ford. *Article on Nomography* Encyclopedia Britannica

35. B. L. Fridman. *Improvement in the smoothness of functions in the Kolmogorov superposition theorem.* Dokl. Akad. Nauk SSSR 177(5), 1019-1022. Soviet Math. Dokl. 8(6)1967, 1550-1553, 1967.

36. H. L. Frisch, C. Borzi, G. Ord, J. K. Perces, G. O. Williams. *Approximate representation of functions of several variables in terms of functions of one variable.* Phys. Rev. Lett. 63(9), 927-929, 1989.

37. F. Girosi, T. Poggio. *Representation properties of Networks: Kolmogorov's theorem is irrelevant.* Neural Computation 1, 465-469, 1989.

38. I. Grattan-Guiness. *A Sideways Look a Hilberts's Twenty-three Problems of 1900.* Notices of the American Mathematical Society, 47(7) 2000.

39. R. Graver. *On the removal of four terms from an equation by means of a Tschirnhaus transformation.* Bull. Amer. Math. Soc. 35, 73-78, 1929.

40. Robert Hecht-Nielsen. *Kolmogorov's mapping neural network existence theorem.* Proc. Inter. Conf. On Neural Networks, Iii, 11-13, IEEE Press, New York, 1987.

41. Robert Hecht-Nielsen. *Neurocomputing.* Eddison-Wesley Pub. Co., New York, 1990.

42. T. H. Hedberg. *The Kolmogorov superposition theorem.* Lecture notes in math. 187, Topics in approximation theory, appendix Ii, Shapiro, H.S., Springer-Verlag, New York, 1971.

43. G. M. Henkin, Vituškin, A.G. (1967) *Linear superpositions of functions.* Russian Math Surveys 22(1), 77-126.

44. David Hilbert. *Mathematische Probleme.* Nach. Akad. Wiss. Göttingen 253-297, 1900; Gesammelte Abhandlungen. 3, 290-329, 1935; *Bull. Amer. Math. Soc.* 8253-297, 1902.

45. David Hilbert. *über die Gleichung neunten Grades,* Math. Ann. 79, 1927 243-250; Gesammelte Abhandlungen. 2(1933), 393-400

46. K. Hornik, M. Stinchomb, H. White. *Multilayer feedforward networks are universal approximations.* Neural networks 2, 359-366, 1989.

47. B. Igelnik, N. Parikh. *Kolmogorov's spline network.* IEEE Transactions on Neural Networks 14(4), 725-733, 2003.

48. J. P. Kahane. *Sur le theoreme de superposition de Kolmogorov.* 44. Approx. Theory 13, 229-234, 1975.

49. J. P. Kahane. *Le treizième problème de Hilbert: un carrefour de l'algèbre, de l'analyse et de la géomètrie.* R. Taton and P. Dugas, Eds.), Cahiers du Sèmin. D. Hostoire d. Math. 3, Inst. H. Poincaré, Paris, 1-25, 1982.

50. J. M. Kantor. *Hilbert's problems and their sequels.* The math. Intelligencer 18(1), 21-30, 1966.

51. H. Katsuura, D. Sprecher. *Computational aspects of Kolmogorov's superposition theorem.* Neural Networks 7(3), 455-461, 1994.

52. J. S. Kim. *Proof of Kolmogorov's theorem.* Master's dissertation, Univ. Maryland, 1960.

53. A. N. Kolmogorov. *Estimation of the minimal number of elements of the epsilon-net in various functional classes and its application to the question of representability of functions of several variables by superpositions of functions of fewer variables.* Dokl. Akad. Nauk SSSR 101, 192-194, 1955.

54. A. N. Kolmogorov. *On the representation of continuous functions of several variables by superpositions of continuous functions of a smaller number of variables.* Dokl. Akad. Nauk SSSR 108, 179-182. Transl. Amer. Math. Soc. 2(17)1961, 369-373, 1956.

55. A. N. Kolmogorov. *On the representation of continuous functions of many variablles by superposition of continuous functions of one variable and addition.* Dokl. Akad. Nauk SSSR 114, 953-956, 1957. Transl. Amer. Math. Soc. 2(28)1963, 55-59,

56. A. N. Kolmogorov. *There is a hypothesis that, in general, Lipschitz continuity or the stronger condition of continuous differentiability excludes the possibility of paradoxical representations of functions of several variables by functions of a smaller number of variables. I am attracted by the following hypothesis: There exists an analytic function of three variables which is not representable by any finite superposition of continuously differentiable functions of two variables; there exist analytic functions of two variables not representable by any finite superposition of continuously differentiable functions of one variable and addition. A special case of the second part of the hypothesis has recently been proved by Vituškin,* Private communication, translated from Russian by G. G. Lorentz, Oct. 18, 1965.

57. M. Köppen. *On the training of a Kolmogorov network.* ICANN 2002, Lecture Notes in Computer Science, 2415 (2002), 474-479.

58. A. S. Kronrod. *On functions of two variables.* Uspehi Mat. Nauk 5(1), 24-134, 1950.

59. V. Kůrková. *13th Hilbert's problem and neural networks.* Theoretical aspects of neural networks, 213-216, World Scientific Publ. Co., Singapore, 1991.

60. V. Kůrková. *Kolmogorov's theorem is relevant.* Neural Computation 3, 617-622, 1991.

61. V. Kůrková. *Kolmogorov's theorem and multilayer neural networks.* Neural Networks 5, 501-506, 1992.

62. P-E. Leni, Y.D. Fougerolle, F. Truchetet. *A novel approach for image sweeping functions using approximating schemes.* Université de Bourgogne, Le Creusot, France.

63. J. Lin, R. Unbehauen. A simplified model for understanding the Kolmogorov network. Preprint, 1-4, 1992.

64. V. Lin. *Around the 13$^{th}$ Hilbert problem for algebraic functions.* DEpt. of Math. Technion 1991.

65. G. G. Lorentz. *Metric entropy, widths, and superpositions of functions.* Amer. Math. Monthly 69(6), 469-485, 1962.

66. G. G. Lorentz. *Entropy and its applications.* Siam Numer. Anal. Ser. B 1, 97-103, 1964.

67. G. G. Lorentz. Approximation of functions. Holt, Rinehart and Winston, N.Y., 1966.

68. G. G. Lorentz. *Metric entropy and approximation.* Bul. Amer. Math. Soc. 72(6), 903-937, 1966.

69. G. G. Lorentz. *On the 13th problem of Hilbert.* Mathematical developments arising from the Hilbert problems. Proc. Amer. Math. Soc. Symposia Pure Math, 28, 419-430, 1976.

70. K. Menger. *Kurventheorie.* Berlin-Leipzig, Chap. X, 1932.

71. A. A. Miljutin. *Isomorphism of the spaces of continuous functions over compact sets of the cardinality of the continuum.* Teor. Funkcii Funkcional Anal. I Prilozen. Vyp. 2, 150-156, 1966.

72. V. P. Motornyi. *On the question of the best approximation of functions of two variables by functions of the form $F(X)+G(Y)$.*Izv. Akad. Nauk SSSR Ser. Mat. 27, 1211-1214, 1963.

73. M. Nakamura, R. Mines, V. Kreinovich. *Guaranteed Intervals for Kolmogorov's Theorem (and Their Possible Relation to Neural Networks).* Interval Computations 3(1993), 183-199.

74. Manuela Nees. *Approximative versions of Kolmogorov's superposition theorem, proved constructively.* J. Comput. Appl. Math. 54, 239-250, 1994.

75. Yu. P. Ofman. *On the best approximation of functions of two variables by functions of the form F(X)+G(Y).* Izv. Akad. Nauk SSSR Ser. Mat. 25, 239-252, 1961.

76. P. A. Ostrand. *Dimension of metric spaces and Hilbert's problem 13.* Bul. Amer. Math. Soc. 71(4), 619-622, 1965.

77. Ostrovski. *Über Dirichletsche Reihen und algebraische Differentialgleichungen,* Math. Zeit. 8, 241-98, 1920.

78. T. Poggio, F. Girosi. Networks for approximation and learning. Proc. IEEE 78, 1481-1497, 1990.

79. G. Pólya G. Szägo. *Aufgaben und Lehrsätze.* Vol. I, Part II (119), Dover, New York, 1945. (Problems 119-119(A), 61-62; Answers 220-223.)

80. W. H. Raudernbush, JR. *On Hilbert's thirteenth Paris problem.* Bull. Amer. Math. Soc. 33(4), 433-34, 1927.

81. C. Reid. *Hilbert.* Springer Verlag 1970.

82. H. Sagan. *Space-Filling Curves.* Springer Verlag, 1994.

83. D. A. Sprecher, A. Bernat, V. Kreinovitch, L. Lomgpré. *Optimal universal and parallel computer, neural networks, and Kolmogorov's theorem.* (unpublished)

84. D. A. Sprecher. *A representation theorem for continuous functions of several variables.* Proc. Amer. Math. Soc. 16(2), 200-203, 1965.

85. D. A. Sprecher. *On the structure of continuous functions of several variables.* Trans. Amer. Math. Soc. 115(3), 340-355, 1965.

86. D. A. Sprecher. *On admissibility in representations of functions of several variables as finite sums of functions of one variable.* Numerical solutions of partial diff. eqs., Bramble, J., Ed., Academic Press, New York, 95-109, 1966.

87. D. A. Sprecher. *On the structure of representations of continuous functions of several variables as finite sums of continuous functions of one variable.* Proc. Amer. Math. Soc. 17(1), 98-105, 1966.

88. D. A. Sprecher. *On best approximations in several variables.* J. Reine und Angewandte Math. 229, 117-130, 1968.

89. D. A. Sprecher. *On similarity in functions of several variables.* Amer. Math. Monthly, 76(6), 627-632, 1969.

90. D. A. Sprecher. *A survey of solved and unsolved problems in superposition of functions.* J. Approximation Theory 6, 123-134, 1972.

91. D. A. Sprecher. *An improvement in the superposition theorem of Kolmogorov.* J. Math. Analysis and Applications 38, 208-213, 1972.

92. D. A. Sprecher. *A universal mapping for Kolmogorov's superposition theorem.* Neural Networks 6, 1089-1094, 1993.

93. D. A. Sprecher. *A numerical construction of a universal function for Kolmogorov's superpositions.* Neural Network World 6(4), 711-718, 1996.

94. D. A. Sprecher. *A numerical implementation of Kolmogorov's superpositions.* Neural Networks 9(5), 765-772, 1996.

95. D. A. Sprecher. *A numerical implementation of Kolmogorov's superpositions II.* Neural Networks 10(3), 447-457, 1997.

96. D. A. Sprecher, S. Draghici. *Space-filling curves and Kolmogorov superposition-based neural networks.* Neural Networks 15, 57-67, 2002.

97. D. A. Sprecher. *Kolmogorov Superpositions: A New Computational Algorithm.* Efficiency and Scalability Methods for Computational Intellect Chapter 11, IGI Global, 2013.

98. D. A. Sprecher. *On computational algorithms for real-valued continuous functions of several variables.* Neural Networks 59, 16-22, 2014.

99. Yaki Sternfeld. *Dimension theory and superpositions of continuous functions.* Israel J. Math. 20(3-4), 300-320, 1975.

100. Yaki Sternfeld. *Dimension, superposition of functions and separation of points in compact metric spaces.* Math. Publication Series, Report 74, Univ. of Haifa, 1983. Israel J. Math 50, 13-53, 1985.

101. Yaki Sternfeld. *Hilbert's 13th problem and dimension.* Geometric Aspects of Functional Analysis: Israel Seminar (GAFA) 1987-88. Springer Lecture Notes in Mathematics, 1989.

102. V. N. Trofimov, L. R. Khariton. *On the error of uniform approximation of functions of two variables by functions of one variable.* Izv. Vyss. Zaved. Mat. 70-73 1979. Soviet Math. 23(2)1979, 71-74, 1979.

103. Tschebotarow. *Über ein algebraisches Problem von Herrn Hilbert.* I Math. Ann. 104, 1931; II Math. Anna. 105, 1931.

104. I. A. Vainstein, M. A. Kreines. *Sequences of functions of the form F[X(X)+Y(Y)]*. Uspehi Mat. Nauk 15(4)94, 123-128, 1960.

105. A. G. Vituškin. *On Hilbert's 13th problem*. Dokl. Akad. Nauk SSSR 95, 701-704, 1954.

106. A. G. Vituškin. *Proof of the existence of analytic functions of several variables not representable by linear superpositions of continuously differentiable functions of fewer variables. Dokl. Akad. Nauk. SSSR* 156(6), 1258-1261, 1964. *Soviet Math. Dokl.* 5, 793-796, 1964.

107. A. G. Vituškin. *Representability of functions by superposition of functions of a smaller number of variables*. Proc. International Congress Math. Moscow, 1966. Transl. Amer. Math. Soc. Ser. 2(86)1966, 101-108, 1970.

108. A. G. Vituškin, G. M. Henkin. *Linear superpositions of functions*. Uspehi Mat. Nauk 22(1)133, 77-124, 1967.

109. A. G. Vituškin. *On the representability of functions by superposition of functions of a smaller number of variables*. (abstract), 1968. Transl. Amer. Math. Soc. Ser. 2(70), 255-257.

110. A. G. Vituškin. *On representation of functions by means of superpositions and related topics*. L'enseignement Mathematique, 2nd Ser. xxiii (3-4), 255-320, 1977.

111. A. Wiman. *Über die anwendung der Tschirnhausen-Transformatio auf die reduktion algebraischer gleichungen*. Nova Acta Soc. Sci. Upsal. 3-8, 1928.

# Index